Springer Series in Operations Research and Financial Engineering

Springer Series in Operations Research and Financial Engineering

Altiok: Performance Analysis of Manufacturing Systems
Birge and Louveaux: Introduction to Stochastic Programming
Bonnans and Shapiro: Perturbation Analysis of Optimization Problems
Bramel, Chen, and Simchi-Levi: The Logic of Logistics: Theory, Algorithms, and Applications for Logistics and Supply Chain Management (second edition)
Dantzig and Thapa: Linear Programming 1: Introduction
Dantzig and Thapa: Linear Programming 2: Theory and Extensions
de Haan and Ferreira: Extreme Value Theory: An Introduction
Drezner (Editor): Facility Location: A Survey of Applications and Methods
Facchinei and Pang: Finite-Dimensional Variational Inequalities and Complementarity Problems, Volume I
Facchinei and Pang: Finite-Dimensional Variational Inequalities and Complementarity Problems, Volume II
Fishman: Discrete-Event Simulation: Modeling, Programming, and Analysis
Fishman: Monte Carlo: Concepts, Algorithms, and Applications
Haas: Stochastic Petri Nets: Modeling, Stability, Simulation
Klamroth: Single-Facility Location Problems with Barriers
Muckstadt: Analysis and Algorithms for Service Parts Supply Chains
Nocedal and Wright: Numerical Optimization
Olson: Decision Aids for Selection Problems
Pinedo: Planning and Scheduling in Manufacturing and Services
Pochet and Wolsey: Production Planning by Mixed Integer Programming
Resnick: Extreme Values, Regular Variation, and Point Processes
Resnick: Heavy Tail Phenomena: Probabilistic and Statistical Modeling
Simchi-Levi: The Logic of Logistics: Theory, Algorithms, and Applications for Logistics and Supply Chain Management
Whitt: Stochastic-Process Limits: An Introduction to Stochastic-Process Limits and Their Applications to Queues
Yao (Editor): Stochastic Modeling and Analysis of Manufacturing Systems
Yao and Zheng: Dynamic Control of Quality in Production-Inventory Systems: Coordination and Optimization
Yeung and Petrosyan: Cooperative Stochastic Differential Games

Jean Bernard Lasserre

Linear and Integer Programming vs Linear Integration and Counting

A Duality Viewpoint

Jean Bernard Lasserre
LAAS-CNRS and Institute of Mathematics
University of Toulouse
LAAS, 7 avenue du Colonel Roche
31077 Toulouse Cédex 4
France
lasserre@laas.fr

ISSN1431-8598
ISBN 978-0-387-09413-7 e-ISBN 978-0-387-09414-4
DOI 10.1007/978-0-387-09414-4
Springer Dordrecht Heidelberg London New York

Library of Congress Control Number: 2009920280

Mathematics Subject Classification (2000): 60J05, 37A30, 28A33, 47A35

Printed on acid-free paper

Springer is part of Springer Science+Business Media (www.springer.com)

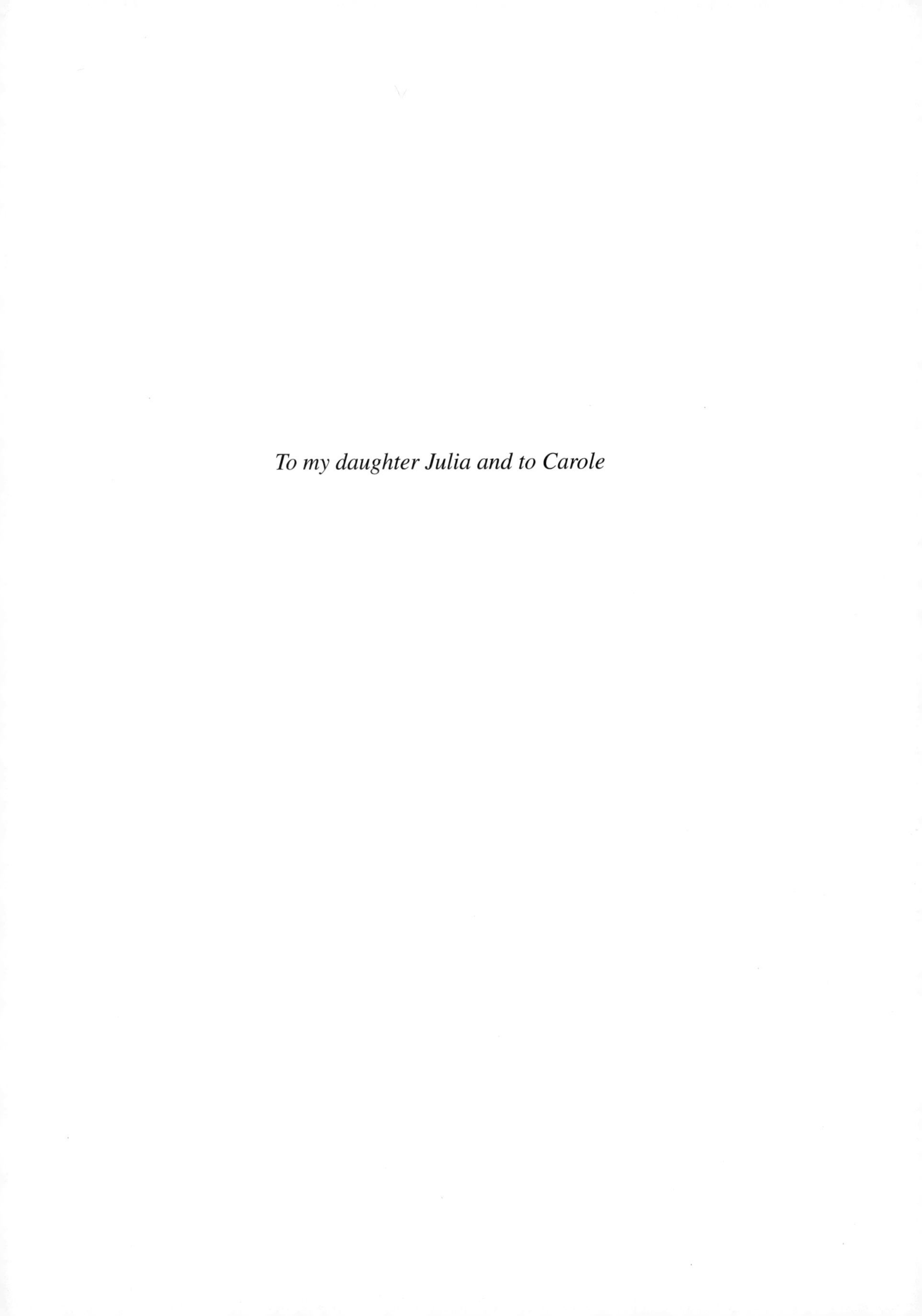

To my daughter Julia and to Carole

Preface

Integer programming (IP) is a fascinating topic. Indeed, while linear programming (LP), its continuous analogue, is well understood and extremely efficient LP software packages exist, solving an integer program can remain a formidable challenge, even for some small size problems. For instance, the following small (5-variable) IP problem (called the *unbounded knapsack problem*)

$$\begin{aligned}
&\min\{213x_1 - 1928x_2 - 11111x_3 - 2345x_4 + 9123x_5\} \\
&\text{s.t. } 12223x_1 + 12224x_2 + 36674x_3 + 61119x_4 + 85569x_5 = 89643482, \\
&\qquad x_1, x_2, x_3, x_4, x_5 \in \mathbb{N},
\end{aligned}$$

taken from a list of difficult knapsack problems in Aardal and Lenstra [2], is not solved even by hours of computing, using for instance the last version of the efficient software package CPLEX.

However, this is *not* a book on integer programming, as very good ones on this topic already exist. For standard references on the theory and practice of integer programming, the interested reader is referred to, e.g., Nemhauser and Wolsey [113], Schrijver [121], Wolsey [136], and the more recent Bertsimas and Weismantel [21]. On the other hand, this book could provide a complement to the above books as it develops a rather unusual viewpoint.

Indeed, one first goal of this book is to analyze and develop some striking analogies between four problems, all related to an integer program $\mathbf{P}_d$ (the subscript "d" being for *discrete*); namely its associated *linear* programming problem $\mathbf{P}$, its associated *linear integration* problem $\mathbf{I}$, and its associated *linear summation* (or *linear counting*) problem $\mathbf{I}_d$. In fact, while $\mathbf{P}$ and $\mathbf{P}_d$ are the respective $(\max,+)$-algebra analogues of $\mathbf{I}$ and $\mathbf{I}_d$, $\mathbf{P}_d$, and $\mathbf{I}_d$ are the respective discrete analogues of $\mathbf{P}$ and $\mathbf{I}$. In addition, the same simple relationship links the value of $\mathbf{P}$ with that of $\mathbf{I}$ on the one hand, and the value of $\mathbf{P}_d$ with that of $\mathbf{I}_d$ on the other hand.

If LP duality is of course well understood (as a special case of Legendre–Fenchel duality in convex analysis), IP duality is much less developed although there is a well-known *superadditive* dual associated with $\mathbf{P}_d$. On the other hand, the linear integration and linear counting problems $\mathbf{I}$ and $\mathbf{I}_d$ have well-defined respective dual problems $\mathbf{I}^*$ and $\mathbf{I}^*_d$, although they are not qualified as such in the literature. Indeed, $\mathbf{I}^*$ (resp., $\mathbf{I}^*_d$) is obtained from the inverse Laplace transform (resp., inverse $\mathbb{Z}$-transform) applied to the Laplace transform (resp., $\mathbb{Z}$-transform) of the value function, exactly in the same way the LP dual $\mathbf{P}^*$ is obtained from the Legendre–Fenchel transform applied to the Legendre–Fenchel transform of the value function. Moreover, recent results by people like

Barvinok, Brion, and Vergne, and Pukhlikov and Khovanskii have provided nice and elegant exact formulas for $\mathbf{I}$ and $\mathbf{I}_d$. One purpose of this book is to show that a careful analysis of these formulas permit us to shed some interesting light on the links and analogies between the (continuous) integration and (discrete) counting programs $\mathbf{I}$ and $\mathbf{I}_d$.

In addition, in view of connections and analogies among $\mathbf{P}, \mathbf{P}_d, \mathbf{I}$, and $\mathbf{I}_d$ on the one hand, and duality results already available for $\mathbf{P}, \mathbf{I}$, and $\mathbf{I}_d$ on the other, another goal is to provide new insights and results on duality for integer programming, and to reinterpret some already existing results in light of these new results and analogies.

This book is an attempt to reach this goal, and throughout all chapters our investigation is guided by these analogies, which are not just formal but rest on a rigorous mathematical analysis. We hope to convince the reader that they are also useful to better understand in particular the difference between the discrete and continuous cases and reasons why the former is significantly more difficult. We also hope that some of the material presented here could be introduced in graduate courses in optimization and operations research, as this new viewpoint makes a link between problems that after all are not so different when looked at through a distinct lens. Indeed, very often the discrete and continuous cases are treated separately (as are integration and optimization) and taught in different courses. The associated research communities are also distinct. On a more practical side, some duality results presented in the last chapters of the book may also provide new ideas for generating efficient *cuts* for integer programs, a crucial issue for improving the powerful software packages already available, like CPLEX and XPRESS-MP codes.

Finally, let us mention that IP is also studied in the algebraic geometry research community, and standard algebraic tools like Gröbner basis, Gröbner fan, and toric ideals permit us to better understand IP as an arithmetic refinement of LP and to re-interpret some known (algebraic) concepts, like Gomory relaxations, for example.

The plan of the book is as follows: We first introduce the four related problems $\mathbf{P}, \mathbf{P}_d, \mathbf{I}$, and $\mathbf{I}_d$ in Chapter 1. Next, in Part I we analyze problem $\mathbf{I}$ and compare with $\mathbf{P}$ in Chapters 2 and 3. In Part II, we do the same with $\mathbf{I}_d$ and $\mathbf{P}_d$ in Chapters 4 and 5. Part III is then devoted to various duality results, including (i) the link between the old concept of Gomory relaxations and exact formula obtained by Brion, and Brion and Vergne, in Chapters 6 and 7, (ii) a discrete Farkas lemma and a characterization of the integer hull in Chapters 8 and 9, and (iii) the link with the superadditive dual in Chapter 10.

Some elementary background on Legendre–Fenchel, Laplace, and $\mathbb{Z}$-transforms, as well as Cauchy residue theorem in complex analysis, can be found in the Appendix.

Acknowledgements

This book benefited from several stimulating discussions with (among others) Alexander Barvinok, Matthias Beck, Jesus de Loera, Jon Lee, Serkan Hosten, Bernd Sturmfels, Rekha Thomas, and Michelle Vergne during several visits and/or workshops at the *Mathematical Sciences Research Institute* (MSRI, Berkeley), the *Fields Institute for Research in Mathematical Sciences* (Toronto), and the *Institute for Mathematics and Its Applications* (IMA, Minneapolis). I gratefully acknowledge financial support from the AMS and the above institutes, which made these visits possible.

Some results presented in this book were obtained in collaboration with Eduardo Santillan Zeron from Cinvestav-IPN (México D.F.) with financial support from a research cooperation program between CNRS (France) and CONACYT (Mexico), which is also gratefully acknowledged.

Finally, I also want to thank the CNRS institution for providing a very pleasant working environment at the LAAS-CNRS laboratory in Toulouse.

Toulouse
24 July 2008

Jean B. Lasserre

Contents

Chapter 1
Introduction

With $A \in \mathbb{Z}^{m \times n}$, $c \in \mathbb{R}^n$, $y \in \mathbb{Z}^m$, consider the *integer program* (IP)

$$\mathbf{P}_d : \max_x \{c'x | Ax = y, x \geq 0, x \in \mathbb{Z}^n\}. \tag{1.1}$$

This discrete analogue of linear programming (LP) is a fundamental NP-hard problem with numerous important applications. However, solving $\mathbf{P}_d$ remains in general a formidable computational challenge, sometimes even for some small size problems. For standard references on the theory and practice of integer programming the interested reader is referred to, e.g., Nemhauser and Wolsey [113], Schrijver [121], Wolsey [136], and Bertsimas and Weismantel [21].

One motivation for this book is to develop *duality concepts* for the integer program $\mathbf{P}_d$, and the way we do it is by analyzing and developing some (not so well-known) striking analogies between the four problems $\mathbf{P}, \mathbf{I}, \mathbf{P}_d$, and $\mathbf{I}_d$ described below. So far, and to the best of our knowledge, most duality results available for integer programs are obtained via the use of *superadditive* functions as in, e.g., Gomory and Johnson [59], Jeroslow [69, 70], Johnson [73, 74], and Wolsey [134, 135], and the smaller class of *Chvátal* and *Gomory* functions as in, e.g., Blair and Jeroslow [23] (see also Schrijver [121, pp. 346–353]). For instance, the following dual problem is associated with the integer program $\mathbf{P}_d$ in (1.1):

$$\mathbf{P}_d^* : \min_f \{f(y) | f(A_j) \geq c_j, j = 1, \ldots, n, f \in \Gamma\}, \tag{1.2}$$

where Γ is the set of functions $f : \mathbb{Z}^m \to \mathbb{R}$ that are *superadditive*, i.e., $f(x+y) \geq f(x) + f(y)$ for all $x, y \in \mathbb{Z}^m$, and such that $f(0) = 0$.

Then, $\mathbf{P}_d^*$ in (1.2) is a problem dual of $\mathbf{P}_d$ in the traditional sense. By this we mean that we retrieve the usual properties of

J.B. Lasserre, *Linear and Integer Programming vs Linear Integration and Counting*,
Springer Series in Operations Research and Financial Engineering, DOI 10.1007/978-0-387-09414-4_1,

- *weak duality*, i.e., $c'x \leq f(y)$ for all primal feasible $0 \leq x \in \mathbb{Z}^n$ and dual feasible $f \in \Gamma$ and
- *strong duality*, i.e., if either $\mathbf{P}_d$ or $\mathbf{P}_d^*$ has a finite optimal value then there exist an optimal primal solution $0 \leq x^* \in \mathbb{Z}^n$ and an optimal dual solution $f^* \in \Gamma$ such that $c'x^* = f^*(y)$.

We come back to this in Chapter 10. However, superadditive Chvátal and Gomory functions are only defined implicitly from their properties, and the resulting dual problems as (1.2) defined in, e.g., [23] or [134] are essentially conceptual in nature. And so, Gomory functions are rather used to generate valid inequalities and cutting planes for the primal problem $\mathbf{P}_d$. An exception is the recent work of Klabjan [80, 81], which addresses (1.2) directly and proposes an algorithm to compute an optimal superadditive function.

We claim that additional *duality* concepts for integer programs can be derived from the $\mathbb{Z}$-transform (or generating function) associated with the *counting* version $\mathbf{I}_d$ (defined below) of the integer program $\mathbf{P}_d$. Results for counting problems, notably by Barvinok [13, 14], Barvinok and Pommersheim [15], Khovanskii and Pukhlikov [79], and in particular, Brion and Vergne's counting formula [27], will prove to be especially useful. But we also claim and show that those duality concepts are in fact the discrete analogues of some already existing duality concepts for two other related "continuous" problems, namely the *linear program* $\mathbf{P}$ (the continuous analogue of $\mathbf{P}_d$) and the *linear integration* problem $\mathbf{I}$ (the continuous analogue of $\mathbf{I}_d$). In fact, so far the notion of "duality" is rather unusual for problems $\mathbf{I}$ and $\mathbf{I}_d$, but this duality will appear very natural later in the book, precisely when one looks at $\mathbf{I}$ and $\mathbf{I}_d$ in a certain light.

1.1 The four problems $\mathbf{P}$,$\mathbf{P}_d$,$\mathbf{I}$,$\mathbf{I}_d$

Given $A \in \mathbb{R}^{m \times n}$(or $\mathbb{Z}^{m \times n}$), $y \in \mathbb{R}^m$(or $\mathbb{Z}^m$), and $c \in \mathbb{R}^n$, let $\Omega(y) \subset \mathbb{R}^n$ be the convex polyhedron

$$\Omega(y) := \{x \in \mathbb{R}^n | Ax = y, \ x \geq 0\},$$

and consider the four related problems $\mathbf{P}, \mathbf{P}_d, \mathbf{I}$, and $\mathbf{I}_d$ displayed in Table 1.1, in which the integer program $\mathbf{P}_d$ appears in the upper right corner (s.t. is the abbreviation for "subject to").

Table 1.1

Linear Programming		Integer Programming
$\mathbf{P}: \ f(y,c) := \max c'x$	$\longleftrightarrow$	$\mathbf{P}_d : f_d(y,c) := \max c'x$
$x \in \Omega(y)$		$x \in \Omega(y) \cap \mathbb{Z}^n$
$\updownarrow$		$\updownarrow$
Linear Integration		**Linear Counting (Summation)**
$\mathbf{I}: \widehat{f}(y,c) := \int_{\Omega(y)} e^{c'x} d\sigma$	$\longleftrightarrow$	$\mathbf{I}_d : \widehat{f}_d(y,c) := \sum_{x \in \Omega(y) \cap \mathbb{Z}^n} e^{c'x}$

Problem $\mathbf{I}$ (in which $d\sigma$ denotes the Lebesgue measure on the affine variety $\{x \in \mathbb{R}^n | Ax = y\}$ that contains the convex polyhedron $\Omega(y)$) is the linear integration version of the linear program $\mathbf{P}$, whereas $\mathbf{I}_d$ is the *linear counting* version of the (discrete) integer program $\mathbf{P}_d$. In addition, $\mathbf{P}_d$ and $\mathbf{I}_d$ are the discrete analogues of $\mathbf{P}$ and $\mathbf{I}$, respectively. We say linear integration (resp., linear counting) because the function to integrate (resp., to sum up) is the exponential of a linear form, whereas the domain of integration (resp., summation) is a polyhedron (resp., the lattice points of a polyhedron).

Why should these four problems help in analyzing $\mathbf{P}_d$? Because, first, $\mathbf{P}$ is related with $\mathbf{I}$ in the same simple manner as $\mathbf{P}_d$ is related with $\mathbf{I}_d$. Next, as we will see, nice results (which we call *duality* results) are available for $\mathbf{P}, \mathbf{I}$, and $\mathbf{I}_d$ and extend in a natural way to $\mathbf{P}_d$.

In fact, $\mathbf{P}$ and $\mathbf{P}_d$ are the respective $(\max,+)$-algebra formal analogues of $\mathbf{I}$ and $\mathbf{I}_d$ in the usual algebra $(+,\times)$. In the $(\max,+)$-algebra the addition $a \oplus b$ stands for $\max(a,b)$ and the $\times$ operation is the usual addition $+$; indeed, the "max" in $\mathbf{P}$ and $\mathbf{P}_d$ can be seen as an *idempotent* integral (or, *Maslov integral*) in this algebra (see, e.g., Litvinov et al. [106]). For a nice parallel between results in probability ($(+,\times)$-algebra) and optimization ($(\max,+)$-algebra), the reader is referred to, e.g., Bacelli et al. [10, §9] and Akian et al. [5]. In other words, one may define a symbolic abstract problem

$$\int_{\Omega(y)\cap\Delta}^{\oplus} e^{c'x}$$

and the situation is depicted in Table 1.2, where the precise meaning of the symbol "$\int^{\oplus}$" depends on whether

(a) the underlying algebra $(\oplus, \odot)$ is $(+,\cdot)$ or $(\max,+)$
(b) the underlying environment is continuous ($\Delta = \mathbb{R}^n$) or discrete ($\Delta = \mathbb{Z}^n$)

Table 1.2 The four problems $\mathbf{P}, \mathbf{P}_d, \mathbf{I}$, and $\mathbf{I}_d$

	$(+,\cdot)$	$(\max,+)$
$\Delta = \mathbb{R}^n$	$\mathbf{I}: \int^{\oplus} = \int$	$\mathbf{P}: \int^{\oplus} = \max$
$\Delta = Z^n$	$\mathbf{I}_d: \quad \int^{\oplus} = \Sigma$	$\mathbf{P}_d: \int^{\oplus} = \max$

Of course, notice that by monotonicity of the exponential,

$$\begin{aligned} e^{f(y,c)} &= e^{\max\{c'x : x \in \Omega(y)\}} = \max\{e^{c'x} : x \in \Omega(y)\}, \\ e^{f_d(y,c)} &= e^{\max\{c'x : x \in \Omega(y) \cap \mathbb{Z}^n\}} = \max\{e^{c'x} : x \in \Omega(y) \cap \mathbb{Z}^n\}, \end{aligned}$$

and so, $\mathbf{P}$ and $\mathbf{I}$, as well as $\mathbf{P}_d$ and $\mathbf{I}_d$, are simply related via

$$e^{f(y,c)} = \lim_{r\to\infty} \widehat{f}(y,rc)^{1/r}, e^{f_d(y,c)} = \lim_{r\to\infty} \widehat{f_d}(y,rc)^{1/r}. \tag{1.3}$$

Equivalently, by continuity of the logarithm,

$$f(y,c) = \lim_{r\to\infty} \frac{1}{r} \ln \widehat{f}(y,rc), \quad f_d(y,c) = \lim_{r\to\infty} \frac{1}{r} \ln \widehat{f_d}(y,rc), \tag{1.4}$$

a relationship that will be useful later. Next, concerning *duality*, the *Legendre–Fenchel* transform

$$f(\cdot,c) \mapsto \mathscr{F}[f(\cdot,c)] = f^*(\cdot,c) : \mathbb{R}^n \to \mathbb{R},$$

which yields the usual dual LP of $\mathbf{P}$,

$$\mathbf{P}^* : f^{**}(y,c) = \mathscr{F}[f^*(\cdot,c)](y) = \min_{\lambda\in\mathbb{R}^m} \{y'\lambda \,|\, A'\lambda \geq c\} = f(y,c), \tag{1.5}$$

has a natural analogue for integration, the *Laplace transform*, and thus, the *inverse Laplace transform* problem (that we call $\mathbf{I}^*$) is the formal analogue of $\mathbf{P}^*$ and provides a nice duality for integration (although not usually presented in these terms in the literature). Finally, the $\mathbb{Z}$-*transform* is the obvious analogue for *summation* of the Laplace transform for integration, and we will see that in the light of recent results for *counting problems*, it permits us to establish a nice duality for $\mathbf{I}_d$ of the same vein as duality for the (continuous) integration problem $\mathbf{I}$. In addition, by (1.4), it also provides a powerful tool to analyze the integer program $\mathbf{P}_d$.

1.2 Summary of content

We first review what we call *duality* principles available for $\mathbf{P}$, $\mathbf{I}$, and $\mathbf{I}_d$ and underline parallels and connections between them. In particular, the parallel between what we call the *duals* of $\mathbf{I}$ and $\mathbf{I}_d$ permits us to highlight a fundamental difference between the continuous and discrete cases. Namely, in the former, the data appear as *coefficients* of the dual variables, whereas in the latter, the *same* data appear as *exponents* of the dual variables (the effect of integration versus summation). Consequently, the (discrete) $\mathbb{Z}$-transform (or generating function) has many more *poles* than the Laplace transform. While the Laplace transform has only *real* poles, the $\mathbb{Z}$-transform has additional *complex* poles associated with each real pole, which induces some *periodic* behavior, a well-known phenomenon in number theory where the $\mathbb{Z}$-transform (or *generating function*) is a standard tool (see, e.g., Iosevich [68], Mitrinovíc et al. [111]). So, if the procedure of inverting the Laplace transform or the $\mathbb{Z}$-transform (i.e., solving the dual problems $\mathbf{I}^*$ and $\mathbf{I}_d^*$) is basically of the same nature, i.e., the (complex) integration of a rational function, it is significantly more complicated in the discrete case, due to the presence of these additional complex poles.

Then we use these results to analyze the discrete optimization problem $\mathbf{P}_d$. Central in the analysis is Brion and Vergne's discrete formula [27] for counting problems. In particular, we provide a closed form expression of the optimal value $f_d(y,c)$, one which highlights the special role played by the so-called *reduced costs* of the linear program $\mathbf{P}$ and the *complex poles* of the $\mathbb{Z}$-transform associated with each basis of the linear program $\mathbf{P}$. We also show that each basis A_σ of the linear program $\mathbf{P}$ provides $\det(A_\sigma)$ complex *dual* vectors in $\mathbb{C}^m$ which are the complex (periodic) analogues for $\mathbf{P}_d$ of the unique dual vector in $\mathbb{R}^m$ for $\mathbf{P}$, associated with the basis A_σ. As in linear programming (but in a more complicated way), the optimal value $f_d(y,c)$ of $\mathbf{P}_d$ can be found by inspection of (certain sums of) reduced costs associated with each vertex of $\Omega(y)$. Finally, both optimal values of $\mathbf{P}_d$ and $\mathbf{P}$ obey a similar formula. In the latter, one formally replaces the complex terms of the former by their modulus.

We next relate Brion and Vergne's discrete formula to the so-called Gomory relaxations introduced in the algebraic approach of Gomory [58] in the late 1960s. In addition, we also present a simple algorithm for $\mathbf{P}_d$, based on Barvinok's counting algorithm, and show how it is also closely related to Gomory relaxations.

We also provide a *discrete* Farkas lemma for the existence of nonnegative integral solutions $x \in \mathbb{N}^n$ to $Ax = y$ and reinterpret the standard Farkas lemma as its continuous analogue. Moreover, this discrete version permits us to derive a linear program $\mathbf{Q}$, equivalent to $\mathbf{P}_d$, and a characterization of the integer hull $\mathrm{conv}(\Omega(y) \cap \mathbb{Z}^n)$ of the polyhedron $\Omega(y)$. Finally, we end up by showing that the LP dual of $\mathbf{Q}$ can be interpreted in terms of the superadditive dual of $\mathbf{P}_d$.

Part I
Linear Integration and Linear Programming

Chapter 2
The Linear Integration Problem I

In the first part of the book, we consider the linear integration problem **I** and establish a comparison with its (max, +)-algebra analogue, the linear programming (LP) problem **P**.

2.1 Introduction

In this chapter, we are interested in the *linear integration* problem **I** defined in Chapter 1, that is,

$$\mathbf{I}: \widehat{f}(y,c) := \int_{\Omega(y)} e^{c'x}\, d\sigma, \tag{2.1}$$

where $\Omega(y) \subset \mathbb{R}^n$ is the convex polyhedron

$$\Omega(y) := \{x \in \mathbb{R}^n | Ax = y, \quad x \geq 0\} \tag{2.2}$$

for some given matrix $A \in \mathbb{R}^{m \times n}$, and vectors $y \in \mathbb{R}^m$, $c \in \mathbb{R}^n$, all with rational entries, and where $d\sigma$ is the Lebesgue measure on the smallest affine variety that contains $\Omega(y)$. When $c \equiv 0$, A is full rank, and $\Omega(y)$ is compact, we thus obtain the $(n-m)$-dimensional volume of $\Omega(y)$. As already mentioned, the linear program **P** is the $(\max,+)$-algebra analogue of problem **I** in the usual algebra $(+,\times)$, and we will see in the next chapter that the analogy between **I** and **P** is not simply formal.

We describe what we call a *dual* algorithm for computing the volume of $\Omega(y)$, i.e., problem **I** with $c \equiv 0$, because it is a standard problem in computational geometry with many interesting and important applications, and also because the same algorithm works when $c \neq 0$, with *ad hoc* and obvious modifications. Also it can be a viable alternative or complement to the various primal methods briefly presented below.

J.B. Lasserre, *Linear and Integer Programming vs Linear Integration and Counting*,
Springer Series in Operations Research and Financial Engineering, DOI 10.1007/978-0-387-09414-4_2,

Some computational complexity results

Computing the volume $\mathrm{vol}(\Omega)$ of a convex polytope Ω (or integrating over Ω) is difficult; its computational complexity is discussed in, e.g., Bollobás [24] and Dyer and Frieze [46]. Indeed, any deterministic algorithm with polynomial time complexity that would compute upper and lower bounds $\overline{v}$ and $\underline{v}$ on $\mathrm{vol}(\Omega)$ cannot yield an upper bound $g(n)$ on $\overline{v}/\underline{v}$ better than a polynomial in the dimension n.

A *convex body* $K \subset \mathbb{R}^n$ is a compact convex subset with nonempty interior. A strong separation *oracle* answers either $x \in K$ or $x \notin K$, and in the latter case produces a hyperplane separating x from K. An algorithm is a sequence of questions to the oracle, with each question depending on the answers to previous questions. The complexity of the algorithm is the number of questions asked before upper and lower bounds $\overline{\mathrm{vol}}(K)$ and $\underline{\mathrm{vol}}(K)$ are produced. Let $B^n \subset \mathbb{R}^n$ be the Euclidean unit ball of $\mathbb{R}^n$. If $r_1 B^n \subset K \subset r_2 B^n$ for some positive numbers r_1, r_2, the algorithm is said to be *well guaranteed*. The input size of a convex body satisfying $r_1 B^n \subset K \subset r_2 B^n$ is $n + \langle r_1 \rangle + \langle r_2 \rangle$, where $\langle x \rangle$ is the number of binary digits of a dyadic rational x.

Theorem 2.1. *For every polynomial time algorithm for computing the volume of a convex body $K \subset \mathbb{R}^n$ given by a well-guaranteed separation oracle, there is a constant $c > 0$ such that*

$$\frac{\overline{\mathrm{vol}}(K)}{\underline{\mathrm{vol}}(K)} \leq \left(\frac{cn}{\log n} \right)^n$$

cannot be guaranteed for $n \geq 2$.

However, Lovász [107] proved that there is a polynomial time algorithm that produces bounds $\overline{\mathrm{vol}}(K)$ and $\underline{\mathrm{vol}}(K)$ satisfying $\overline{\mathrm{vol}}(K)/\underline{\mathrm{vol}}(K) \leq n^n (n+1)^{n/2}$, whereas Elekes [54] proved that for $0 < \varepsilon < 2$ there is no polynomial time algorithm that produces $\overline{\mathrm{vol}}(K)$ and $\underline{\mathrm{vol}}(K)$ with $\overline{\mathrm{vol}}(K)/\underline{\mathrm{vol}}(K) \leq (2-\varepsilon)^n$.

In contrast with this negative result, and if one accepts randomized algorithms that fail with small probability, then the situation is much better. Indeed, the celebrated Dyer, Frieze, and Kanan's probabilistic approximation algorithm [47] computes the volume to fixed arbitrary relative precision ε, in time polynomial in ε^{-1}. The latter algorithm uses approximation schemes based on rapidly mixing Markov chains and isoperimetric inequalities.

More precisely, following [24], let $K \subset \mathbb{R}^n$ with $n \geq 2$, and let ε, η be small positive numbers. An ε-approximation of $\mathrm{vol}(K)$ is a number $\widehat{\mathrm{vol}}(K)$ such that

$$(1-\varepsilon)\,\widehat{\mathrm{vol}}(K) \leq \mathrm{vol}(K) \leq (1+\varepsilon)\,\widehat{\mathrm{vol}}(K). \tag{2.3}$$

A *fully polynomial randomized approximation scheme* (FPRAS) is a randomized algorithm that runs in time polynomial in the input size of K, ε^{-1}, $\log \eta^{-1}$, and with probability at least $1-\eta$ produces an ε-approximation (2.3) of $\mathrm{vol}(K)$. Then the important result of Dyer, Frieze, and Kanan [47] states that there exists a FPRAS for the volume of a convex body given by a well-guaranteed membership oracle.

Exact methods

Basically, methods for *exact* computation of the volume (triangulations or simplicial decompositions) can be classified according to whether one has a *half-space* description

$$\Omega = \{x \in \mathbb{R}^n | Ax \leq y\}$$

or a *vertex* description

$$\Omega = \left\{ x \in \mathbb{R}^n | x = \sum_{j=1}^{p} \lambda_j x(j), \lambda_j \geq 0, \sum_{j=1}^{p} \lambda_j \leq 1 \right\}$$

of Ω, or when both descriptions are available. For instance, Delaunay's triangulation (see, e.g., [29]) and von Hohenbalken's simplicial decomposition [131] both require the list of vertices, whereas Lasserre's algorithm [94] requires a half-space description. On the other hand, Lawrence's [102] and Brion and Vergne's [27] formulas, as well as Cohen and Hickey's triangulation method [34], require both half-space and vertex descriptions of the polytope. On the other hand, Barvinok's algorithm [13] computes the volume by computing the integral of $e^{c'x}$ over Ω for a small c, i.e., evaluates the Laplace transform of the function $x \mapsto \mathrm{I}_\Omega(x)$ at the particular $\lambda = c$. We call these approaches *primal* because they all work directly in the primal space $\mathbb{R}^n$ of the variables describing the polytope regardless of whether Ω has a vertex or half-space description.

In this chapter, we take a *dual*[1] approach, that is, we consider problem **I** as that of evaluating the value function $\widehat{f}(y,0)$ at some particular $y \in \mathbb{R}^m$, and to do so, we compute the inverse Laplace transform of its Laplace transform $\widehat{F}(\lambda,0)$ at the point $y \in \mathbb{R}^m$; we call this the *dual integration problem* $\mathbf{I}^*$. In the present context, as $\widehat{F}$ is available in closed form, computing the inverse Laplace transform at the point y is simply the evaluation of a complex integral. This method is *dual* in nature because contrast to primal methods which work in the primal space $\mathbb{R}^n$ of the x variables, we instead work in the space $\mathbb{R}^m$ of *dual* variables λ associated with the nontrivial constraints $Ax = y$. In summary:

- Triangulations or signed decomposition methods, as well as Barvinok's algorithm, are primal methods that work in the space $\mathbb{R}^n$ of primal variables, regardless of whether Ω has a vertex or half-space description. The right-hand side y is fixed.
- Our dual type algorithm works in $\mathbb{C}^m$ as it uses residue techniques to *invert* the Laplace transform $\widehat{F}(\lambda,0) : \mathbb{C}^m \to \mathbb{C}$ of the function $f(\cdot,0) : \mathbb{R}^m \to \mathbb{R}$, at the particular point $y \in \mathbb{R}^m$. So $c = 0$ is fixed, y varies as the argument of $\widehat{f}(\cdot,0)$, and λ varies in the inverse Laplace integral.

2.2 Primal methods

Primal methods for exact volume computation can be divided into two classes: triangulations and signed decompositions. A basic result used in all these methods is the exact formula for the volume of a simplex in $\mathbb{R}^n$. Let $\Delta(x_0,\ldots,x_n)$ be the simplex of $\mathbb{R}^n$ with vertices $x_0,\ldots,x_n \in \mathbb{R}^n$. Then

$$\mathrm{vol}\,(\Delta(x_0,\ldots,x_n)) = \frac{|\det(x_1 - x_0, x_2 - x_0, \ldots, x_n - x_0)|}{n!}. \tag{2.4}$$

[1] Duality here has nothing to do with duality between the *vertex* and *half-space* descriptions of a convex polytope $\Omega \subset \mathbb{R}^n$.

Triangulations

A *triangulation* of an n-polytope $\Omega \subset \mathbb{R}^n$ is a set $\{\Delta_i\}_{i=1}^s$ of n-simplices $\Delta_i \subset \mathbb{R}^n$ such that no two distinct simplices have an interior point in common. And so,

$$\mathrm{vol}\,(\Omega) = \sum_{i=1}^{s} \mathrm{vol}\,(\Delta_i),$$

with $\mathrm{vol}\,(\Delta_i)$ given by (2.4). Hence, most of the work is concentrated in finding a triangulation of Ω. For instance, let $\Omega \subset \mathbb{R}^2$ be the rectangle $ABDC$ in Figure 2.1. With E inside Ω, one has the triangulation $(AEC),(AEB),(BED),(DCEC)$ of Ω. Among the triangulation methods are Delaunay's triangulation, boundary triangulation, and Cohen and Hickey's triangulation [34]. While the first two only require a vertex representation of Ω, the latter requires both vertex and half-space descriptions.

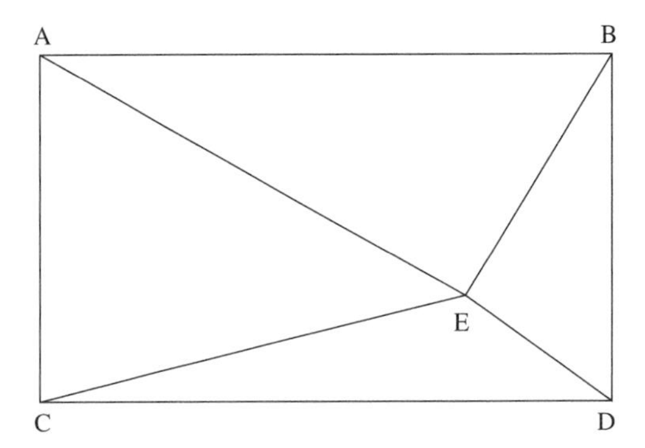

FIGURE 2.1 Triangulation of a rectangle

Delaunay's triangulation is obtained by lifting the polytope Ω onto the surface of an $(n+1)-$ dimensional convex body, by $x \mapsto (x, f(x))$ for some strictly convex function $f : \mathbb{R}^n \rightarrow \mathbb{R}$. The lifted vertices $(v, f(v))$ must be in the general position (i.e., their convex hull simplicial). Then, interpreting the facets in terms of the original vertices yields the desired triangulation.

Boundary triangulation for simplicial polytopes (i.e., each facet is an $(n-1)$-dimensional simplex) links each facet with an interior point as in Figure 2.1 so as to yield the desired triangulations. For general polytopes, a small perturbation yields a simplicial polytope. Cohen and Hickey's triangulation is an improvement of boundary triangulation as it considers a vertex instead of an interior point.

Signed decompositions

As another primal method, one may decompose Ω into *signed* n-simplices $\{\Delta_i\}$ whose signed union is exactly Ω. That is, one writes $\Omega = \bigcup_i \varepsilon_i \Delta_i$ with $\varepsilon_i \in \{-1,+1\}$, and where the signed decomposition $\{\varepsilon_i \Delta_i\}$ must satisfy the following requirement for every point x not on the boundary of any simplex: If $x \in \Omega$ then x appears in exactly one more positive Δ_i, whereas if $x \notin \Omega$ then it must appear equally often in positive and negative Δ_i. And so, we now have

$$\mathrm{vol}(\Omega) = \sum_{i=1}^{s} \varepsilon_i \,\mathrm{vol}(\Delta_i).$$

In Figure 2.2 the same rectangle Ω has the signed decomposition

$$[+(AB,AC,e),+(e,CD,BD),-(AB,e,BD),-(AC,CD,e)],$$

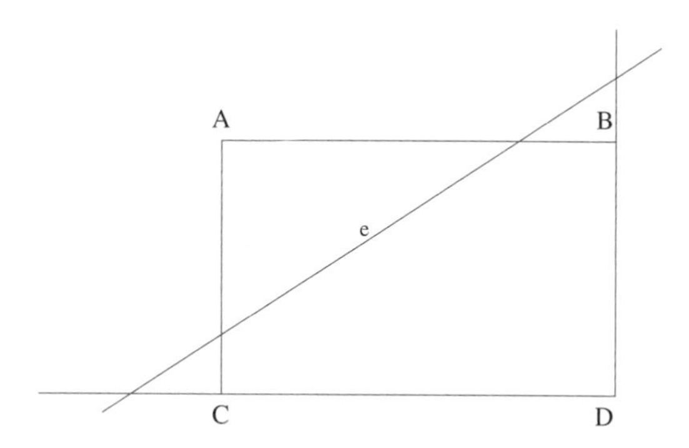

FIGURE 2.2 Signed decomposition of a rectangle

where a triplet (a,b,c) stands for the unique simplex determined by the intersection of lines a,b, and c. Filliman's duality [56] shows that triangulations and signed decompositions are in duality via the facet–vertex duality (i.e., to a vertex corresponds a facet and viceversa). Namely, let $\Omega \subset \mathbb{R}^n$ be a polytope with the origin in its interior and let T be a triangulation of Ω such that the origin is not in the union of hyperplanes defining the triangulation T. Then T induces a signed decomposition of the polar

$$\Omega^\circ = \{x \in \mathbb{R}^n | \langle x,y\rangle \leq 1, \forall y \in \Omega\}$$

of Ω, in such a way that each simplex $\Delta(a_0,\ldots,a_n)$ is in T if and only if the unique bounded simplex determined by the corresponding hyperplanes in the polar Ω° is in the signed decomposition. The sign of a simplex in the signed decomposition is determined by its separation parity in the triangulation.

Among signed decomposition methods, let us cite Lawrence's decomposition [102], which results in a simple and elegant formula. With $A \in \mathbb{R}^{m\times n}$, let $\Omega := \{x \in \mathbb{R}^n | Ax \leq y\}$ be simple, and for a vertex v of Ω, let $A_v \in \mathbb{R}^{n\times n}$ be the submatrix of A formed with the rows defining the binding constraints at v. Let $c \in \mathbb{R}^n$ and $q \in \mathbb{R}$ be such that $c'x+q$ is not constant along any edge, and let $\pi^v := c'A_v^{-1}$ be well defined. Then

$$\mathrm{vol}(\Omega) = \frac{(c'x+q)^n}{n!|\det A_v| \prod_{i=1}^n \pi_i^v}. \tag{2.5}$$

We will see in Section 2.3 that (2.5) is the same as (2.16) obtained from Brion and Vergne's continuous formula (2.15) (given for Ω described in standard equality form (2.2)).

Alternatively, let $\mathrm{vol}_i(n-1,\Omega_i)$ be the $(n-1)$-dimensional volume of the facet $\Omega_i := \Omega \cap \{x : (Ax)_i = y_i\}$. Then the formula

$$n\,\mathrm{vol}\,(\Omega) = \sum_{i=1}^{m} \frac{y_i}{\|A_i\|} \mathrm{vol}(n-1,\Omega_i), \tag{2.6}$$

obtained in Lasserre [94] as just Euler's identity for homogeneous functions (applied to $\mathrm{vol}\,(\Omega)$ as a function of the right-hand side y), has the well-known equivalent geometric formulation

$$n\,\mathrm{vol}\,(\Omega) = \sum_{i=1}^{m} d(O,H_i)\,\mathrm{vol}(n-1,\Omega_i)$$

(where $d(O,H_i)$ is the algebraic distance from the origin to the hyperplane $A_i x = y_i$), the n-dimensional analogue of the two-dimensional formula (height)(base)/2 for the triangle area. From this geometric interpretation, (2.6) is also a signed decomposition of Ω when O is outside Ω, and a triangulation when O is inside Ω. For instance, if E in Figure 2.1 is the origin and Ω_i is the segment (CD), then $y_i/\|A_i\|$ is the distance from E to (CD) and $y_i\mathrm{vol}\,(1,\Omega_i)/2\|A_i\|$ is just the classical formula for the area of the triangle ECD. Formula (2.6) permits us to derive a recursive algorithm as one iterates and one applies again this formula for the polytope Ω_i, for all $i=1,\ldots,m$ (see Lasserre [94]).

Concerning the computational complexity and efficiency of triangulation and signed decomposition methods, the interested reader is referred to Büeler et al. [29]. In their computational study, they have observed that triangulation methods behave well for near-simplicial polytopes, whereas signed decomposition methods behave well for near-simple polytopes. This opposite behavior is typical for the hypercube and its polar, the cross polytope. Also, Delaunay's triangulation as well as Lawrence's formula exhibited serious numerical instability in many examples. The latter is due to difficulties in finding a *good* vector c.

Barvinok's algorithm

For each vertex v of $\Omega \subset \mathbb{R}^n$ let $K_v := \mathrm{co}\,(\Omega, v)$ be the tangent cone (or supporting cone) of Ω at v, that is,

$$\mathrm{co}\,(\Omega, v) = \{x \in \mathbb{R}^n | \varepsilon x + (1-\varepsilon)v \in \Omega, \forall \text{ sufficiently small } \varepsilon > 0\}. \tag{2.7}$$

Then Brion [25] proved that

$$s(\Omega) = \int_{\Omega} \mathrm{e}^{c'x}\,dx = \sum_{v:\,\text{vertex of }\Omega} \int_{K_v} \mathrm{e}^{c'x}\,dx =: \sum_{v:\,\text{vertex of }\Omega} s(K_v).$$

In Barvinok's algorithm [13], each cone K_v is the union $\bigcup_j K_{vj}$ of *simple* cones $\{K_{vj}\}$ so that $s(K_v) = \sum_j s(K_{vj})$. In addition, let K_v be given as the convex hull of its extreme rays $\{u_1,\ldots,u_m\} \subset \mathbb{R}^n$. Then

given $c \in \mathbb{C}^m$, one can compute $s(K_v)$ in $O(n^3 m \binom{m}{n})$ arithmetic operations. Finally, taking c very small yields an approximation of the volume of Ω.
As already mentioned, the above methods are primal as they all work in the primal space $\mathbb{R}^n$ of primal variables, regardless of whether Ω has a vertex or half-space representation. We next consider a method that works in the dual space $\mathbb{R}^m$ of variables associated with the nontrivial constraints $Ax = y$ in the description (2.2) of $\Omega(y)$.

2.3 A dual approach

Let $y \in \mathbb{R}^m$ and $A \in \mathbb{R}^{m \times n}$ be real-valued matrices and consider the convex polyhedron $\Omega(y)$ defined in (2.2). Assume that

$$\{x \in \mathbb{R}^n | Ax = 0; c'x \geq 0, x \geq 0\} = \{0\}. \tag{2.8}$$

That is, infinite lines in $\Omega(y)$ with $c'x \geq 0$ are excluded, which makes sense as we want to integrate $e^{c'x}$ over $\Omega(y)$.

Remark 2.1. By a special version of Farkas' lemma due to Carver (see, e.g., Schrijver [121, (33), p. 95]) and adapted to the present context, (2.8) holds if and only if there exists $u \in \mathbb{R}^m$ such that $A'u > c$.

The Laplace transform

Consider the function $\widehat{f}(\cdot, c) : \mathbb{R}^m \to \mathbb{R}$ defined by

$$y \mapsto \widehat{f}(y, c) := \int_{\Omega(y)} e^{c'x} d\sigma, \tag{2.9}$$

where $d\sigma$ is the Lebesgue measure on the smallest affine variety that contains $\Omega(y)$.

Let $\widehat{F}(\cdot, c) : \mathbb{C}^m \to \mathbb{C}$ be its m-dimensional two-sided Laplace transform, i.e.,

$$\lambda \mapsto \widehat{F}(\lambda, c) := \int_{\mathbb{R}^m} e^{-\langle \lambda, y \rangle} \widehat{f}(y, c)\, dy \tag{2.10}$$

(see, e.g., [28], and Appendix).

Theorem 2.2. *Let $\Omega(y)$ be the convex polytope in (2.2) and let $\widehat{f}$ and $\widehat{F}$ be as in (2.9) and (2.10), respectively. Then*

$$\widehat{F}(\lambda, c) = \frac{1}{\prod_{j=1}^{n} (A'\lambda - c)_j}, \qquad \Re(A'\lambda - c) > 0. \tag{2.11}$$

Moreover,

$$\widehat{f}(y,c) = \frac{1}{(2\pi i)^m} \int_{\gamma_1 - i\infty}^{\gamma_1 + i\infty} \cdots \int_{\gamma_m - i\infty}^{\gamma_m + i\infty} e^{\langle \lambda, y \rangle} \widehat{F}(\lambda)\, d\lambda \tag{2.12}$$

where $\gamma \in \mathbb{R}^m$ *satisfies* $A'\gamma > c$.

Proof. Apply the definition (2.10) of $\widehat{F}$, to obtain

$$\begin{aligned}
\widehat{F}(\lambda) &= \int_{\mathbb{R}^m} e^{-\langle \lambda, y \rangle} \left[\int_{x \geq 0,\, Ax = y} e^{c'x} d\sigma \right] dy \\
&= \int_{\mathbb{R}^n_+} e^{\langle c - A'\lambda, x \rangle} dx \\
&= \prod_{j=1}^{n} \left[\int_0^\infty e^{(c - A'\lambda)_j x_j} dx_i \right] \\
&= \frac{1}{\prod_{j=1}^n (A'\lambda - c)_j}, \quad \text{with } \Re(A'\lambda - c) > 0.
\end{aligned}$$

And (2.12) is obtained by a direct application of the inverse Laplace transform; see [28] and Appendix. It remains to show that the domain $\{\lambda \in \mathbb{C}^m : \Re(A'\lambda - c) > 0\}$ is nonempty. However, this follows from Remark 2.1. □

The dual problem $\mathbf{I}^*$

We define the dual problem $\mathbf{I}^*$ to be the *inversion* problem:

$$\begin{aligned}
\mathbf{I}^* : \widehat{f}(y,c) &:= \frac{1}{(2i\pi)^m} \int_{\gamma_m - i\infty}^{\gamma_1 + i\infty} \cdots \int_{\gamma_m - i\infty}^{\gamma_m + i\infty} e^{\langle y, \lambda \rangle} \widehat{F}(\lambda, c)\, d\lambda \\
&= \frac{1}{(2i\pi)^m} \int_{\gamma_m - i\infty}^{\gamma_1 + i\infty} \cdots \int_{\gamma_m - i\infty}^{\gamma_m + i\infty} \frac{e^{\langle y, \lambda \rangle}}{\prod_{j=1}^n (A'\lambda - c)_j} d\lambda,
\end{aligned} \tag{2.13}$$

where $\gamma \in \mathbb{R}^m$ is fixed and satisfies $A'\gamma - c > 0$.

We may indeed call $\mathbf{I}^*$ a dual problem of $\mathbf{I}$ as it is the inversion of the Laplace transform of $\widehat{f}$, exactly like $\mathbf{P}^*$ is the inversion of the Legendre–Fenchel transform of f. It is defined on the space $\mathbb{C}^m$ of variables $\{\lambda_j\}$ associated with the nontrivial constraints $Ax = y$; notice that we also retrieve the standard "ingredients" of the dual optimization problem $\mathbf{P}^*$, namely, $y'\lambda$ and $A'\lambda - c$. Incidentally, observe that the domain of definition (3.7) of $\widehat{F}(\cdot, c)$ is precisely the interior of the feasible set of the dual problem $\mathbf{P}^*$ in (3.5). We will comment more on this when we compare problems $\mathbf{P}$ and $\mathbf{I}$ in Section 3.2.

As mentioned earlier, computing $\widehat{f}(y,c)$ at the point $y \in \mathbb{R}^m$ via (2.12), i.e., by solving the dual problem $\mathbf{I}^*$ (2.13), reduces to computing the Laplace inverse of $\widehat{F}(\lambda,c)$ at the point y. For instance, this complex integral can be evaluated by repeated application of Cauchy residue theorem. That is, one computes the integral (2.12) by successive one-dimensional complex integrals with respect to (w.r.t.) one variable λ_k at a time (e.g., starting with $\lambda_1, \lambda_2, \ldots$) and by repeated application of Cauchy residue theorem. This is possible because the integrand is a rational fraction, and after application of Cauchy residue theorem at step k w.r.t. λ_k, the output is still a rational fraction of the remaining variables $\lambda_{k+1}, \ldots, \lambda_m$; see Section 2.4. It is not difficult to see that the whole procedure is a summation of partial results, each of them corresponding to a (multipole) vector $\widehat{\lambda} \in \mathbb{R}^m$ that annihilates m terms of n products in the denominator of the integrand. This is formalized in the nice formula of Brion and Vergne [27, Proposition 3.3, p. 820], which we describe below.

Brion and Vergne's continuous formula

A remarkable result of Brion and Vergne provides us with an *explicit* expression of $\widehat{f}(y,c)$. However, direct evaluation of this formula may pose some numerical instability, and in Section 4.4 we also provide a residue algorithm to evaluate $\widehat{f}(y,c)$ at a point $y \in \mathbb{R}^m$.

Before stating Brion and Vergne's formula, we first introduce some notation. Write the matrix $A \in \mathbb{R}^{m \times n}$ as $A = [A_1|\ldots|A_n]$ where $A_j \in \mathbb{R}^m$ denotes the jth column of A for all $j = 1, \ldots, n$. With $\Delta := (A_1, \ldots, A_n)$ let $C(\Delta) \subset \mathbb{R}^m$ be the closed convex cone generated by Δ.

A subset σ of $\{1, \ldots, n\}$ is called a *basis* of Δ if the sequence $\{A_j\}_{j \in \sigma}$ is a basis of $\mathbb{R}^m$, and the set of bases of Δ is denoted by $\mathscr{B}(\Delta)$. For $\sigma \in \mathscr{B}(\Delta)$ let $C(\sigma)$ be the cone generated by $\{A_j\}_{j \in \sigma}$. With any $y \in C(\Delta)$ associate the intersection of all cones $C(\sigma)$ that contain y. This defines a subdivision of $C(\Delta)$ into polyhedral cones. The interiors of the maximal cones in this subdivision are called *chambers*; see, e.g., Brion and Vergne [27]. For every $y \in \gamma$, the convex polyhedron $\Omega(y)$ in (2.2) is *simple*. Next, for a chamber γ (whose closure is denoted by $\overline{\gamma}$), let $\mathscr{B}(\Delta, \gamma)$ be the set of bases σ such that γ is contained in $C(\sigma)$, and let $\mu(\sigma)$ denote the volume of the convex polytope $\{\sum_{j \in \sigma} t_j A_j \,|\, 0 \leq t_j \leq 1\}$ (that is, the determinant of the square matrix $[A_{j \in \sigma}]$). Observe that for $y \in \overline{\gamma}$ and $\sigma \in \mathscr{B}(\Delta, \gamma)$ we have $y = \sum_{j \in \sigma} x_j(\sigma) A_j$ for some $x_j(\sigma) \geq 0$. Therefore, the vector $x(\sigma) \in \mathbb{R}^n_+$ with $x_j(\sigma) = 0$ whenever $j \notin \sigma$ is a *vertex* of the polytope $\Omega(y)$. In the LP terminology, the bases $\sigma \in \mathscr{B}(\Delta, \gamma)$ correspond to *feasible bases* of the linear program $\mathbf{P}$. Denote by V the subspace $\{x \in \mathbb{R}^n \,|\, Ax = 0\}$. Finally, given $\sigma \in \mathscr{B}(\Delta)$, let $\pi_\sigma \in \mathbb{R}^m$ be the row vector that solves $\pi_\sigma A_j = c_j$ for all $j \in \sigma$. A vector $c \in \mathbb{R}^n$ is said to be *regular* if $c_j - \pi_\sigma A_j \neq 0$ for all $\sigma \in \mathscr{B}(\Delta)$ and all $j \notin \sigma$.

Let $c \in \mathbb{R}^n$ be regular with $-c$ in the interior of the dual cone $(\mathbb{R}^n_+ \cap V)^*$ (which is the case if $A'u > c$ for some $u \in \mathbb{R}^m$). Then, Brion and Vergne's formula [27, Proposition 3.3, p. 820] states that

$$\widehat{f}(y,c) = \sum_{\sigma \in \mathscr{B}(\Delta,\gamma)} \frac{e^{\langle c, x(\sigma) \rangle}}{\mu(\sigma) \prod_{k \notin \sigma} (-c_k + \pi_\sigma A_k)} \qquad \forall y \in \overline{\gamma}. \tag{2.14}$$

Notice that in the linear programming terminology, $c_k - \pi_\sigma A_k$ is nothing less than the so-called *reduced cost* of the variable x_k, w.r.t. the basis $\{A_j\}_{j\in\sigma}$. When $y \in \gamma$, the polytope $\Omega(y)$ is *simple* and we can rewrite (2.14) as

$$\widehat{f}(y,c) = \sum_{x(\sigma):\text{ vertex of }\Omega(y)} \frac{e^{\langle c,x(\sigma)\rangle}}{\mu(\sigma)\prod_{j\notin\sigma}(-c_j + \pi_\sigma A_j)}. \tag{2.15}$$

On the other hand, when $y \in \overline{\gamma}\setminus\gamma$ for several chambers γ, then in the summation (2.14), several bases σ correspond to the same vertex $x(\sigma)$ of $\Omega(y)$.

Remarkably, evaluating $\widehat{f}(\cdot,c)$ at a point $y \in \mathbb{R}^m$ reduces to sum up over all vertices $x(\sigma)$ of $\Omega(y)$, the quantity $e^{c'x(\sigma)} B_\sigma^{-1}$, where B_σ is just the product of all reduced costs multiplied by the determinant of the matrix basis A_σ. However, this formula is correct only when c is regular. So, for instance, if one wishes to compute the volume $\widehat{f}(y,0)$ of $\Omega(y)$, i.e., with $c = 0$, one cannot apply directly formula (2.15). However, one also has

$$\mathrm{vol}\,(\Omega(y)) = \frac{\langle -c,x(\sigma)\rangle^n}{n!\,\mu(\sigma)\prod_{j\notin\sigma}(c_j - \pi_\sigma A_j)}. \tag{2.16}$$

In fact, (2.16) is exactly Lawrence's formula (2.5) when the polytope $\Omega(y)$ is in standard equality form (2.2). As mentioned already, this formula is not stable numerically, because despite a regular vector $c \in \mathbb{R}^n$ being generic, in many examples the product in the denominator can be very small and one ends up adding and subtracting huge numbers; see, e.g., the empirical study in Büeler et al. [29].

2.4 A residue algorithm for problem I*

We have just seen that evaluating $\widehat{f}(\cdot,c)$ at a point $y \in \mathbb{R}^m$ reduces to evaluating formula (2.15). However, this formula is correct only when c is regular. Moreover, even if c is regular it may pose some numerical difficulties. The same is true for the volume formula (2.16). In this section we provide a residue algorithm that handles the case where c is not regular; when $c = 0$ the algorithm returns the volume of the convex polytope $\Omega(y)$.

As computing the volume of a convex polyhedron is a basic problem of interest in computational geometry, we will specify the algorithm for the case $c = 0$ and for the inequality case description of $\Omega(y)$, i.e., when

$$\Omega(y) := \{x \in \mathbb{R}^n | Ax \le y, x \ge 0\},$$

in which case an easy calculation yields

$$\widehat{F}(\lambda,c) = \frac{1}{\prod_{k=1}^{m} \lambda_k \prod_{j=1}^{n} (A'\lambda - c)_j}, \qquad \Re(\lambda), \Re(A'\lambda - c) > 0. \tag{2.17}$$

But it will become apparent that the case $c \neq 0$ can be handled exactly in the same manner. Therefore, we wish to evaluate the complex integral

$$\text{volume}\,(\Omega(y)) = \frac{1}{(2\pi i)^m} \int_{\gamma_1 - i\infty}^{\gamma_1 + i\infty} \cdots \int_{\gamma_m - i\infty}^{\gamma_m + i\infty} e^{y'\lambda}\, \widehat{F}(\lambda,0)\, d\lambda,$$

with $\widehat{F}(y,c)$ as in (2.17) and where the real vector $0 < \gamma \in \mathbb{R}^m$ satisfies $A'\gamma > 0$. To make calculations a little easier, with $y \in \mathbb{R}^m$ fixed, we restate the problem as that of computing the value $h(1)$ of the function $h: \mathbb{R} \to \mathbb{R}$ given by

$$z \mapsto h(z) := \widehat{f}(zy,0) = \frac{1}{(2\pi i)^m} \int_{\gamma_1 - i\infty}^{\gamma_1 + i\infty} \cdots \int_{\gamma_m - i\infty}^{\gamma_m + i\infty} e^{z\langle \lambda, y\rangle}\, \widehat{F}(\lambda,0)\, d\lambda. \tag{2.18}$$

Computing the complex integral (2.18) can be done in two ways that are explored below. We do it directly in Section 2.4 by integrating first w.r.t. λ_1, then w.r.t. λ_2, λ_3, ...; or indirectly in Section 2.4, by first computing the one-dimensional Laplace transform H of h and then computing the Laplace inverse of H.

The direct method

To better understand the methodology behind the direct method and for illustration purposes, consider the case of a convex polytope $\Omega(y)$ with only two ($m = 2$) nontrivial constraints.

The m = 2 example. Let $A \in \mathbb{R}^{2 \times n}$ be such that $x = 0$ is the only solution of $\{x \geq 0, Ax \leq 0\}$ so that $\Omega(y)$ is a convex polytope (see Remark 2.1). Moreover, suppose that $y_1 = y_2 = z$ and $A' := [a\,|\,b]$ with $a, b \in \mathbb{R}^n$. For ease of exposition, assume that

- $a_j b_j \neq 0$ and $a_j \neq b_j$ for all $j = 1, 2, \ldots, n$.
- $a_j / b_j \neq a_k / b_k$ for all $j, k = 1, 2, \ldots, n$.

Observe that these assumptions are satisfied with probability equal to one if we add to every coefficient a_i, b_i a perturbation ε, randomly generated under a uniform distribution on $[0, \bar{\varepsilon}]$, with $\bar{\varepsilon}$ very small. Then

$$\widehat{F}(\lambda,0) = \frac{1}{\lambda_1 \lambda_2} \times \frac{1}{\prod_{j=1}^{n} (a_j \lambda_1 + b_j \lambda_2)}, \qquad \begin{cases} \Re(\lambda) > 0 \\ \Re(a\lambda_1 + b\lambda_2) > 0\,. \end{cases}$$

Next, fix γ_1 and $\gamma_2 > 0$ such that $a_j \gamma_1 + b_j \gamma_2 > 0$ for every $j = 1, 2, \ldots n$ and compute the integral (2.18) as follows. We first evaluate the integral

$$I_1(\lambda_2) := \frac{1}{2\pi i} \int_{\gamma_1 - i\infty}^{\gamma_1 + i\infty} \frac{e^{z\lambda_1}}{\lambda_1 \prod_{j=1}^{n} (a_j \lambda_1 + \lambda_2 b_j)}\, d\lambda_1, \tag{2.19}$$

using classical Cauchy residue technique. That is, we (a) close the path of integration adding a semicircle Γ of radius R large enough, (b) evaluate the integral using Cauchy residue theorem (see Appendix), and (c) show that the integral along Γ converges to zero when $R \to \infty$.

Now, since we are integrating w.r.t. λ_1 and we want to evaluate $h(z)$ at $z = 1$, we must add the semicircle Γ on the left side of the integration path $\{\Re(\lambda_1) = \gamma_1\}$ because $e^{z\lambda_1}$ converges to zero when $\Re(\lambda_1) \to -\infty$. Therefore, we must consider *only* the poles of $\lambda_1 \mapsto \widehat{F}(\lambda_1, \lambda_2)$ whose real part is strictly less than γ_1 (with λ_2 being fixed). Recall that $\Re(-\lambda_2 b_j/a_j) = -\gamma_2 b_j/a_j < \gamma_1$ whenever $a_j > 0$, for each $j = 1, 2, \ldots, n$, and $\widehat{F}(\lambda_1, \lambda_2)$ has only poles of first order (with λ_2 being fixed). Then, the evaluation of (2.19) follows easily, and

$$I_1(\lambda_2) = \frac{1}{\lambda_2^n \prod_{j=1}^n b_j} + \sum_{a_j>0} \frac{-e^{-(b_j/a_j)z\lambda_2}}{b_j \lambda_2^n \prod_{k\neq j}(-a_k b_j/a_j + b_k)}.$$

Therefore,

$$\begin{aligned} h(z) &= \frac{1}{2\pi i} \int_{\gamma_2 - i\infty}^{\gamma_2 + i\infty} \frac{e^{z\lambda_2}}{\lambda_2} I_1(\lambda_2)\, d\lambda_2 \\ &= \frac{1}{2\pi i} \int_{\gamma_2 - i\infty}^{\gamma_2 + i\infty} \frac{e^{z\lambda_2}}{\lambda_2^{n+1} \prod_{j=1}^n b_j}\, d\lambda_2 \\ &\quad - \sum_{a_j>0} \frac{1}{2\pi i} \int_{\gamma_2 - i\infty}^{\gamma_2 + i\infty} \frac{a_j^n\, e^{(1-b_j/a_j)z\lambda_2}}{\lambda_2^{n+1} a_j b_j \prod_{k\neq j}(b_k a_j - a_k b_j)}\, d\lambda_2. \end{aligned}$$

These integrals must be evaluated according to whether $(1 - b_j/a_j)y$ is positive or negative, or, equivalently (as $a_j > 0$), whether $a_j > b_j$ or $a_j < b_j$. Thus, recalling that $z > 0$, each integral is equal to

- its residue at the pole $\lambda_2 = 0 < \gamma_2$ when $a_j > b_j$, and
- zero if $a_j < b_j$ because there is no pole on the right side of $\{\Re(\lambda_2) = \gamma_2\}$.

That is,

$$h(z) = \frac{z^n}{n!} \left[\frac{1}{\prod_{j=1}^n b_j} - \sum_{b_j<a_j} \frac{(a_j - b_j)^n}{a_j b_j \prod_{k\neq j}(b_k a_j - a_k b_j)} \right]. \tag{2.20}$$

Observe that the formula is not symmetric in the parameters a, b. This is because we have chosen to integrate first w.r.t. λ_1; and the set $\{j \,|\, b_j < a_j\}$ is different from $\{j \,|\, a_j < b_j\}$, which would have been considered had we integrated first w.r.t. λ_2. In the latter case, we would have obtained

$$h(z) = \frac{z^n}{n!} \left[\frac{1}{\prod_{j=1}^n a_j} - \sum_{a_j<b_j} \frac{(b_j - a_j)^n}{a_j b_j \prod_{k\neq j}(a_k b_j - b_k a_j)} \right], \tag{2.21}$$

which is (2.20), interchanging a and b. Moreover, moving terms around, we get for free the following identity:

$$\sum_{j=1}^n \frac{(a_j - b_j)^n}{a_j b_j \prod_{k\neq j}(b_k a_j - a_k b_j)} = \frac{1}{\prod_{j=1}^n b_j} - \frac{1}{\prod_{j=1}^n a_j}. \tag{2.22}$$

The direct method algorithm. The above methodology easily extends to an arbitrary number m of nontrivial constraints. One evaluates the integral of the right hand side of (2.18) by integrating first w.r.t. λ_1, then w.r.t. λ_2, and so on. The resulting algorithm can be described with a tree of depth $m+1$ ($m+1$ "levels").

Direct method algorithm: Let $0 < c \in \mathbb{R}^m$ be such that $A'c > 0$.

- Level 0 is the root of the tree.
- Level 1 is the integration w.r.t. λ_1 and consists of at most $n+1$ nodes associated with the poles $\lambda_1 := \rho_j^1$, $j = 1,\ldots,n+1$, of the rational function $\prod_i \lambda_i^{-1} \prod_j (A'\lambda)_j^{-1}$, seen as a function of λ_1 only. By the assumption on c, there is no pole ρ_j^1 on the line $\Re(\lambda_1) = \gamma_1$. By the Cauchy residue theorem, only the poles at the left side of the integration path $\{\Re(\lambda_1) = \gamma_1\}$, say ρ_j^1, $j \in I_1$, are selected.
- Level 2 is the integration w.r.t. λ_2. After integration w.r.t. λ_1, *each* of the poles ρ_j^1, $j \in I_1$, generates a rational function of $\lambda_2, \lambda_3, \ldots, \lambda_m$ times an exponential, which, seen as a function of λ_2 only, has at most $n+1$ poles $\rho_i^2(j)$, $i = 1,\ldots,n+1$, $j \in I_1$. Thus, level 2 has at most $(n+1)^2$ nodes associated with the poles $\rho_i^2(j)$. Assuming no pole $\rho_i^2(j)$ on the line $\Re(\lambda_2) = \gamma_2$, by the Cauchy residue theorem, only the poles $\rho_i^2(j)$, $(j,i) \in I_2$, located on the correct side of the integration path $\{\Re(\lambda_2) = \gamma_2\}$ are selected. By "correct" side we mean the right side if the coefficient in the exponential is negative, and the left side otherwise.
- Level k, $k \leq m$, consists of at most $(n+1)^k$ nodes associated with the poles $\rho_s^k(i_1, i_2, \ldots, i_{k-1})$, $(i_1, i_2, \ldots, i_{k-1}) \in I_{k-1}$, $s = 1,\ldots,n+1$, of some rational functions of $\lambda_k, \ldots, \lambda_m$, seen as functions of λ_k only. Assuming no pole on the line $\Re(\lambda_k) = \gamma_k$, only the poles $\rho_{i_k}^k(i_1, i_2, \ldots, i_{k-1})$, $(i_1, i_2, \ldots, i_k) \in I_k$, located on the correct side of the integration path $\{\Re(\lambda_k) = \gamma_k\}$ are selected. And so on.
  The last level m consists of at most $(n+1)^m$ nodes and the integration w.r.t. λ_m is trivial as it amounts to evaluating integrals of the form

$$(2\pi i)^{-1} \int_{\gamma_m - i\infty}^{\gamma_m + i\infty} A \lambda_m^{-(n+1)} e^{\alpha z \lambda_m} d\lambda_m$$

  for some coefficients A, α, which yield $A(\alpha z)^n/n!$ for those $\alpha > 0$. Summing up over all nodes of the last level (analogously, all leaves of the tree) provides the desired value.

Computational complexity. Only simple elementary arithmetic operations are needed to compute the nodes at each level, as in Gauss elimination for solving linear systems. Moreover, finding a vector $\gamma \in \mathbb{R}^m$ that satisfies $\gamma > 0$, $A'\gamma > 0$, can be done in time polynomial in the input size of the problem. Therefore, the computational complexity is easily seen to be essentially described by n^m.

However, some care must be taken with the choice of the integration paths as we assume that at each level k there is no pole on the integration path $\{\Re(\lambda_k) = \gamma_k\}$. This issue is discussed later. The algorithm is illustrated on the following simple example with $n = 2, m = 3$.

Example 2.1. Let $e_3 = (1,1,1)$ be the unit vector of $\mathbb{R}^3$ and let $\Omega(ze_3) \subset \mathbb{R}^2$ be the polytope

$$\Omega(ze_3) := \{x \in \mathbb{R}^2_+ | x_1 + x_2 \leq z; -2x_1 + 2x_2 \leq z; 2x_1 - x_2 \leq z\},$$

whose area is $17z^2/48$.

Choose $\gamma_1 = 3$, $\gamma_2 = 2$, and $\gamma_3 = 1$ so that $\gamma_1 > 2\gamma_2 - 2\gamma_3$ and $\gamma_1 > \gamma_3 - 2\gamma_2$.

$$h(z) = \frac{1}{(2\pi i)^3} \int_{\gamma_1 - i\infty}^{\gamma_1 + i\infty} \cdots \int_{\gamma_3 - i\infty}^{\gamma_3 + i\infty} e^{(\lambda_1 + \lambda_2 + \lambda_3)z} \widehat{F}(\lambda)\, d\lambda,$$

with

$$\widehat{F}(\lambda, 0) = \frac{1}{\lambda_1 \lambda_2 \lambda_3 (\lambda_1 - 2\lambda_2 + 2\lambda_3)(\lambda_1 + 2\lambda_2 - \lambda_3)}.$$

Integrate first w.r.t. λ_1; that is, evaluate the residues at the poles $\lambda_1 = 0$, $\lambda_1 = 2\lambda_2 - 2\lambda_3$, and $\lambda_1 = \lambda_3 - 2\lambda_2$ because $0 < z$, $0 < \gamma_1$, $\Re(2\lambda_2 - 2\lambda_3) < \gamma_1$, and $\Re(\lambda_3 - 2\lambda_2) < \gamma_1$. We obtain

$$h(z) = \frac{1}{(2\pi i)^2} \int_{\gamma_2 - i\infty}^{\gamma_2 + i\infty} \int_{\gamma_3 - i\infty}^{\gamma_3 + i\infty} (I_2 + I_3 + I_4)\, d\lambda_2\, d\lambda_3,$$

where

$$I_2(\lambda_2, \lambda_3) = \frac{-e^{(\lambda_2 + \lambda_3)z}}{2\lambda_2 \lambda_3 (\lambda_3 - \lambda_2)(\lambda_3 - 2\lambda_2)},$$

$$I_3(\lambda_2, \lambda_3) = \frac{e^{(3\lambda_2 - \lambda_3)z}}{6\lambda_2 \lambda_3 (\lambda_3 - \lambda_2)(\lambda_3 - 4\lambda_2/3)},$$

$$I_4(\lambda_2, \lambda_3) = \frac{e^{(2\lambda_3 - \lambda_2)z}}{3\lambda_2 \lambda_3 (\lambda_3 - 2\lambda_2)(\lambda_3 - 4\lambda_2/3)}.$$

Next, integrate $I_2(\lambda_2, \lambda_3)$ w.r.t. λ_3. We must consider the poles on the left side of $\{\Re(\lambda_3) = 1\}$, that is, the pole $\lambda_3 = 0$ because $\Re(\lambda_2) = 2$. Thus, we get $-e^{z\lambda_2}/4\lambda_2^3$, and the next integration w.r.t. λ_2 yields $-z^2/8$.

When integrating $I_3(\lambda_2, \lambda_3)$ w.r.t. λ_3, we have to consider the poles $\lambda_3 = \lambda_2$ and $\lambda_3 = 4\lambda_2/3$ on the right side of $\{\Re(\lambda_3) = 1\}$; and we get

$$\frac{-1}{\lambda_2^3} \left[-\frac{e^{2z\lambda_2}}{2} + \frac{3e^{z\lambda_2 5/3}}{8} \right].$$

Recall that the path of integration has a negative orientation, so we have to consider the negative value of residues. The next integration w.r.t. λ_2 yields $z^2(1 - 25/48)$.

Finally, when integrating $I_4(\lambda_2, \lambda_3)$ w.r.t. λ_3, we must consider only the pole $\lambda_3 = 0$, and we get $e^{-z\lambda_2}/8\lambda_2^3$; the next integration w.r.t. λ_2 yields zero. Hence, adding up the above three partial results yields

$$h(z) = z^2 \left[\frac{-1}{8} + 1 - \frac{25}{48} \right] = \frac{17z^2}{48},$$

which is the desired result.

Multiple poles. In the description of the algorithm we have considered the case in which all the poles are *simple* at each step. This is indeed the *generic* case. However, multiple poles may occur at some of the first $m - 1$ one-dimensional integrations. For instance, at the kth step (after integration w.r.t. $\lambda_1, \lambda_2, \ldots, \lambda_{k-1}$), the integrand could have a denominator that includes one (or several) product(s) of the form

$$(a_k\lambda_k + \cdots + a_m\lambda_m) \times (b_k\lambda_k + \cdots + b_m\lambda_m)$$

with

$$\frac{a_k}{b_k} = \frac{a_{k+1}}{b_{k+1}} = \cdots = \frac{a_m}{b_m}; \tag{2.23}$$

i.e., there is a pole of order greater than or equal to 2 w.r.t. λ_k.

This is a *pathological* case, which happens with probability zero on a sample of problems with randomly generated data. Moreover, one may handle this situation in one of two ways:

- *Directly*, by detecting multiple poles and applying an adequate Cauchy residue technique. This procedure requires some extra work (*detection* of (2.23) + *derivation* in Cauchy residue theorem), which is a polynomial in the dimension n. Observe that detection of (2.23) requires some ε-tolerance as it deals with real numbers.
- *Indirectly*, by preprocessing (perturbing) the nonzero elements of the initial matrix A. Let us say $A(i,j) \rightarrow A(i,j) + \varepsilon(i,j)$ with independent identically distributed (i.i.d.) random variables $\{\varepsilon(i,j)\}$ taking values in some interval $[-\overline{\varepsilon}, \overline{\varepsilon}]$ for $\overline{\varepsilon}$ very small.

While the former technique requires extra work, the latter is straightforward, but only provides (with probability equal to one) an approximation of the exact volume.

Paths of integration. In choosing the integration paths $\{\Re(\lambda_k) = \gamma_k\}$, $k = 1, \ldots, m$, we must determine a vector $0 < \gamma \in \mathbb{R}^m$ such that $A'\gamma > 0$. However, this may not be enough when we want to evaluate the integral (2.18) by repeated applications of the Cauchy residue theorem. Indeed, in the tree description of the algorithm, we have seen that at each level $k > 1$ of the tree (integration w.r.t λ_k), one *assumes* that there is *no* pole on the integration path $\{\Re(\lambda_k) = \gamma_k\}$.

For instance, had we set $\gamma_1 = \gamma_2 = \gamma_3 = 1$ (instead of $\gamma_1 = 3$, $\gamma_2 = 2$, and $\gamma_1 = 1$) in the above example, we could not use Cauchy residue theorem to integrate I_2 or I_3 because we would have the pole $\lambda_2 = \lambda_3$ exactly on the path of integration (recall that $\Re(\lambda_2) = \Re(\lambda_3) = 1$); fortunately, this case is *pathological* as it happens with probability zero in a set of problems with randomly generated data $A \in \mathbb{R}^{m\times n}$, and therefore this issue could be ignored in practice. However, for the sake of mathematical rigor, in addition to the constraints $\gamma > 0$ and $A'\gamma > 0$, the vector $\gamma \in \mathbb{R}^m$ must satisfy additional constraints to avoid the above-mentioned pathological problem. We next describe one way to proceed to ensure that c satisfies these additional constraints.

We have described the algorithm as a tree of depth m (level i being the integration w.r.t. λ_i), where each node has at most $n+1$ descendants (one descendant for each pole on the correct side of the integration path $\{\Re(\lambda_i) = \gamma_i\}$). The volume is then the summation of all partial results obtained at each leaf of the tree (that is, each node of level m). We next describe how to "perturb" *on-line* the initial vector $c \in \mathbb{R}^m$ if at some level k of the algorithm there is a pole on the corresponding integration path $\{\Re(\lambda_k) = \gamma_k\}$.

Step 1. Integration w.r.t. λ_1. Choose a real vector $\gamma := (\gamma_1^1, \ldots, \gamma_m^1) > 0$ such that $A'\gamma > 0$ and integrate (2.18) along the line $\Re(\lambda_1) = \gamma_1^1$. From the Cauchy residue theorem, this is done by selecting the (at most $n+1$) poles $\lambda_1 := \rho_j^1$, $j \in I_1$, located on the left side of the integration path $\{\Re(\lambda_1) = \gamma_1^1\}$. Each pole $\rho_j^1, j = 1, \ldots, n+1$ (with $\rho_j^1 := 0$) is a linear combination

$\beta_{j2}^{(1)}\lambda_2+\cdots+\beta_{jm}^{(1)}\lambda_m$ with real coefficients $\{\beta_{jk}^{(1)}\}$, because A is a real-valued matrix. Observe that by the initial choice of γ,

$$\delta_1 := \min_{j=1,\dots,n+1} |\gamma_1^1 - \sum_{k=2}^m \beta_{jk}^{(1)}\gamma_k^1| > 0.$$

Step 2. Integration w.r.t. λ_2. For each of the poles ρ_j^1, $j\in I_1$, selected at step 1, and after integration w.r.t. λ_1, we now have to consider a rational function of λ_2 with at most $n+1$ poles $\lambda_2 := \rho_i^2(j) := \sum_{k=3}^m \beta_{ik}^{(2)}(j)\lambda_k$, $i=1,\dots,n+1$. If

$$\delta_2 := \min_{j\in I_1}\min_{i=1,\dots,n+1}\left|\gamma_2^1 - \sum_{k=3}^m \beta_{ik}^{(2)}(j)\gamma_k^1\right| > 0,$$

then integrate w.r.t. λ_2 on the line $\Re(\lambda_2)=\gamma_2^1$. Otherwise, if $\delta_2 = 0$ we set $\gamma_2^2 := \gamma_2^1+\varepsilon_2$ and $\gamma_k^2 := \gamma_k^1$ for all $k\neq 2$, by choosing $\varepsilon_2>0$ small enough to ensure that

(a) $A'\gamma^2 > 0$

(b) $\delta_2 := \min_{j\in I_1}\min_{i=1,\dots,n+1}\left|\gamma_2^2 - \sum_{k=3}^m \beta_{ik}^{(2)}(j)\gamma_k^2\right| > 0$

(c) $\max_{j=1,\dots,n+1} |\beta_{j2}^{(1)}\varepsilon_2| < \delta_1$.

The condition (a) is basic whereas (b) ensures that there is no pole on the integration path $\{\Re(\lambda_2) = \gamma_2^2\}$. Moreover, what has been done in step 1 remains valid because from (c), $\gamma_1^2 - \sum_{k=2}^m \beta_{jk}^{(1)}\gamma_k^2$ has the same sign as $\gamma_1^1 - \sum_{k=2}^m \beta_{jk}^{(1)}\gamma_k^1$, and, therefore, none of the poles ρ_j^1 has crossed the integration path $\{\Re(\lambda_1)=\gamma_1^1=\gamma_1^2\}$, that is, the set I_1 is unchanged.

Now integrate w.r.t. λ_2 on the line $\Re(\lambda_2)=\gamma_2^2$, which is done via the Cauchy residue theorem by selecting the (at most $(n+1)^2$) poles $\rho_i^2(j)$, $(j,i)\in I_2$, located at the left or the right of the line $\Re(\lambda_2)=\gamma_2^2$, depending on the sign of the coefficient of the argument in the exponential.

Step 3. Integration w.r.t. λ_3. Likewise, for each of the poles $\rho_i^2(j)$, $(j,i)\in I_2$, selected at step 2, we now have to consider a rational function of λ_3 with at most $n+1$ poles $\rho_s^3(j,i) := \sum_{k=4}^m \beta_{sk}^{(3)}(j,i)\lambda_k$, $s=1,\dots,n+1$. If

$$\delta_3 := \min_{(j,i)\in I_2}\min_{s=1,\dots,n+1}\left|\gamma_3^2 - \sum_{k=4}^m \beta_{sk}^{(3)}(j,i)\gamma_k^2\right| > 0,$$

then integrate w.r.t. λ_3 on the line $\Re(\lambda_3)=\gamma_3^2$. Otherwise, if $\delta_3=0$, set $\gamma_3^3 := \gamma_3^2+\varepsilon_3$ and $\gamma_k^3 := \gamma_k^2$ for all $k\neq 3$ by choosing $\varepsilon_3>0$ small enough to ensure that

(a) $A'\gamma^3 > 0$

(b) $\delta_3 := \min_{(j,i)\in I_2}\min_{s=1,\dots,n+1}\left|\gamma_3^3 - \sum_{k=4}^m \beta_{sk}^{(3)}(j,i)\gamma_k^3\right| > 0$

(c) $\max_{j\in I_1}\max_{i=1,\dots,n+1} |\beta_{i3}^{(2)}(j)\varepsilon_3| < \delta_2$

(d) $\max_{j=1,\dots,n+1} |\beta_{j2}^{(1)}\varepsilon_2 + \beta_{j3}^{(1)}\varepsilon_3| < \delta_1$.

As in previous steps, condition (a) is basic. The condition (b) ensures that there is no pole on the integration path $\{\Re(\lambda_3) = \gamma_3^3\}$. Condition (c) (resp., (d)) ensures that none of the poles $\rho_i^2(j)$ considered at step 2 (resp., none of the poles ρ_j^1 considered at step 1) has crossed the line $\Re(\lambda_2) = \gamma_2^3 = \gamma_2^2$ (resp., the line $\Re(\lambda_1) = \gamma_1^3 = \gamma_1^1$). That is, both sets I_1 and I_2 are unchanged.

Next, integrate w.r.t. λ_3 on the line $\Re(\lambda_3) = \gamma_3^3$, which is done by selecting the (at most $(n+1)^3$) poles $\rho_s(j,i)$, $(j,i,s) \in I_3$, located at the left or right of the line $\{\Re(\lambda_3) = \gamma_3^3\}$, depending on the sign of the argument in the exponential. And so on.

Remark 2.2. (i) It is important to notice that ε_k and γ_k^k play no (numerical) role in the integration itself. They are only used to (1) ensure the absence of a pole on the integration path $\{\Re(\lambda_k) = \gamma_k^k\}$ and (2) locate the poles on the left or the right of the integration path. Their numerical value (which can be very small) has no influence on the computation.

(ii) If one ignores the perturbation technique for the generic case, then a depth-first search of the tree requires a polynomial storage space, whereas an exponential amount of memory is needed with the perturbation technique.

The associated transform algorithm

We now describe an alternative to the direct method, one that permits us (a) to avoid evaluating integrals of exponential functions in (2.18) and (b) to avoid making the *on-line* changes of the integration paths described earlier for handling possible (pathological) poles on the integration path.

Recall that we want to compute (2.18) where $y \neq 0$ and $\gamma > 0$ are real vectors in $\mathbb{R}^m$ with $A'\gamma > 0$. It is easy to deduce from (2.1) that $\widehat{f}(y,0) = 0$ whenever $y \leq 0$, so we may suppose without loss of generality that $y_m > 0$; we make the following simple change of variables as well. Let $p = \langle \lambda, y \rangle$ and $d = \langle \gamma, y \rangle$, so $\lambda_m = (p - \sum_{j=1}^{m-1} y_j \lambda_j)/y_m$ and

$$h(z) = \frac{1}{(2\pi i)^m} \int_{\gamma_1 - i\infty}^{\gamma_1 + i\infty} \cdots \int_{\gamma_{m-1} - i\infty}^{\gamma_{m-1} + i\infty} \left[\int_{d-i\infty}^{d+i\infty} e^{zp} \widehat{G} \, dp \right] d\lambda_1 \ldots d\lambda_{m-1},$$

where

$$\widehat{G}(\lambda_1, \ldots, \lambda_{m-1}, p) = \frac{1}{y_m} \widehat{F} \left(\lambda_1, \ldots, \lambda_{m-1}, \frac{p}{y_m} - \sum_{j=1}^{m-1} \frac{y_j}{y_m} \lambda_j \right). \tag{2.24}$$

We can rewrite $h(z)$ as follows:

$$h(z) = \frac{1}{2\pi i} \int_{d-i\infty}^{d+i\infty} e^{zp} H(p) dp, \tag{2.25}$$

with

$$H(p) := \frac{1}{(2\pi i)^{m-1}} \int_{\gamma_1 - i\infty}^{\gamma_1 + i\infty} \cdots \int_{\gamma_{m-1} - i\infty}^{\gamma_{m-1} + i\infty} \widehat{G} \, d\lambda_1 \ldots d\lambda_{m-1}. \tag{2.26}$$

Notice that $H(p)$ is the Laplace transform of $h(z) = \widehat{f}(zy, 0)$, called the *associated transform* of $\widehat{F}(\lambda)$. In addition, since $h(z)$ is proportional to z^n (volume is proportional to the nth power of length),

$$H(p) = C/p^{n+1},$$

where the constant $C = h(1)n!$ gives us the value $h(1) = \widehat{f}(y,0)$ we are looking for. From here, we need only to compute $H(p)$ in (2.26) in order to know the final result. We can do so via repeated application of the Cauchy residue theorem (as in the direct method algorithm of Section 2.4).

Moreover, after each one of the successive $m-1$ one-dimensional integrations calculated in (2.26), we *always* get an *analytic* function. This fact can be proved as follows. Consider for some time the change of variables $s_j = 1/\lambda_j$ for all $j = 1,2,\ldots,m-1$ in (2.26). This change of variables is well-defined because no λ_j is equal to zero, and we obtain

$$H(p) = \frac{1}{(2\pi i)^{m-1}} \int_{|2\gamma_1 s_1 - 1|=1} \cdots \int_{|2\gamma_{m-1} s_{m-1} - 1|=1} R\, ds_1 \ldots ds_{m-1},$$

where $R(s_1,\ldots,s_{m-1},p)$ is a rational function (recall that G and $\widehat{G}$ are both rational) continuous on the Cartesian product of integration paths

$$E = \{|2\gamma_1 s_1 - 1| = 1\} \times \cdots \times \{|2\gamma_{m-1} s_{m-1} - 1| = 1\} \times \{\Re(p) = d\}.$$

Recall that $\widehat{F}(\lambda)$ is defined on the domain $\{\Re(\lambda) > 0, \Re(A'\lambda) > 0\}$, and so R is analytic on the domain $D \subset \mathbb{C}^m$ of all points $(s_1,\ldots,s_{m-1},p) \in \mathbb{C}^m$ such that $|2\beta_j s_j - 1| = 1$, $s_j \neq 0$, $j = 1,2,\ldots,m-1$, and $\Re(p) = \langle \beta, y\rangle$ for some vector $\beta > 0$ in $\mathbb{R}^m$ with $A'\beta > 0$.

Observe that $E \subset D \cup \{0\}$ and each integration path $\{|2\gamma_j s_j - 1| = 1\}$ is a compact circle, $j = 1,\ldots,m-1$. Therefore, after integration w.r.t. s_1, s_2, ..., s_{m-1}, we get that $H(p)$ is *analytic* on the domain of all points $p \in \mathbb{C}$ such that $\Re(p) = \langle \beta, b\rangle$ for some real vector $\beta = (\gamma_1,\ldots,\gamma_{m-1},\beta_m) > 0$ with $A'\beta > 0$.

Similarly, if we only integrate w.r.t. s_1, s_2, ..., s_k, $1 \leq k \leq m-2$, we get a function which is *analytic* on the domain of all points $(s_{k+1},\ldots,s_{m-1},p) \in \mathbb{C}^{m-k}$ such that $|2\beta_j s_j - 1| = 1$, $s_j \neq 0$, $j = k+1,\ldots,m-1$, and $\Re(p) = \langle \beta, y\rangle$ for some real vector $\beta = (\gamma_1,\ldots,\gamma_k,\beta_{k+1},\ldots,\beta_m) > 0$ with $A'\beta > 0$.

Thus, there is no pole on the next integration path $\{|2\gamma_{k+1}s_{k+1} - 1| = 1\}$ or the final one $\{\Re(p) = d\}$. Referring back to the original variables $\{\lambda_k\}$ and (2.26), we translate this fact as follows. After integrating (2.26) w.r.t. λ_1, λ_2, ..., λ_k, any *apparent* pole placed on the next integration path $\{\Re(\lambda_{k+1}) = \gamma_{k+1}\}$ or the final one $\{\Re(p) = d\}$ is a *removable singularity* (see [36, p. 103]). In other words, poles on the next integration path can be removed by *adding together all terms with poles there*. The general method is illustrated with the same example of two nontrivial constraints ($m = 2$) already considered at the beginning of Section 2.4.

The m = 2 example. Recall that $A' := [a\,|\,b]$ with $a, b \in \mathbb{R}^n$ so that

$$\widehat{F}(\lambda,0) = \frac{1}{\lambda_1\lambda_2} \times \frac{1}{\prod_{j=1}^n (a_j\lambda_1 + b_j\lambda_2)}, \qquad \Re(\lambda) > 0, \Re(A'\lambda) > 0$$

and fix $y_1 = y_2 = z$. To compare with the direct method, and as in the beginning of Section 2.4, assume that $a_j b_j \neq 0$ for all $j = 1,\ldots,n$ and $a_j/b_j \neq a_k/b_k$ for all $j \neq k$.

Fix $\lambda_2 = p - \lambda_1$ and choose real constants $\gamma_1 > 0$ and $\gamma_2 > 0$ such that $a_j\gamma_1 + b_j\gamma_2 > 0$ for every $j = 1,2,\ldots,n$. Notice that $\Re(p) = \gamma_1 + \gamma_2$. We obtain $H(p)$ by integrating $\widehat{F}(\lambda_1, p - \lambda_1, 0)$ on the line $\Re(\lambda_1) = \gamma_1$, which yields

$$H(p) = \frac{1}{2\pi i} \int_{\gamma_1 - i\infty}^{\gamma_1 + i\infty} \frac{1}{\lambda_1(p-\lambda_1)} \times \frac{1}{\prod_{j=1}^{n}((a_j - b_j)\lambda_1 + b_j p)} \, d\lambda_1.$$

Next, we need to determine which poles of $\widehat{F}(\lambda_1, p - \lambda_1)$ are on the left (right) side of the integration path $\{\Re(\lambda_1) = \gamma_1\}$ in order to apply the Cauchy residue theorem. Let $J_+ = \{j | a_j > b_j\}$, $J_0 = \{j | a_j = b_j\}$, and $J_- = \{j | a_j < b_j\}$. Then, the poles on the left side of $\{\Re(\lambda_1) = \gamma_1\}$ are $\lambda_1 = 0$ and $\lambda_1 = -b_j p/(a_j - b_j)$ for all $j \in J_+$ because $-b_j \Re(p)/(a_j - b_j) < \gamma_1$. Additionally, the poles on the right side of $\{\Re(\lambda_1) = \gamma_1\}$ are $\lambda_1 = p$ and $\lambda_1 = -b_j p/(a_j - b_j)$ for all $j \in J_-$. Finally, notice that $\widehat{F}(\lambda_1, p - \lambda_1, 0)$ has only poles of first order.

Hence, computing the residues of poles on the left side of $\{\Re(\lambda_1) = \gamma_1\}$ yields

$$H(p) = \frac{1}{\prod_{j \in J_0} p b_j} \left[\frac{1}{p \prod_{j \notin J_0} p b_j} + \right.$$
$$\left. + \sum_{j \in J_+} \frac{-(a_j - b_j)^{n - |J_0|}}{p^2 a_j b_j \prod_{k \notin J_0, k \neq j} (-p b_j a_k + p a_j b_k)} \right].$$

Hence, after moving terms around, we obtain

$$H(p) = \frac{1}{p^{n+1}} \left[\frac{1}{\prod_{j=1}^{n} b_j} - \sum_{a_j > b_j} \frac{(a_j - b_j)^n}{a_j b_j \prod_{k \neq j} (a_j b_k - b_j a_k)} \right].$$

Notice that previous equation holds even for the case $J_0 \neq \emptyset$. Finally, after integration w.r.t. p, we get

$$h(z) = \frac{z}{n!} \left[\frac{1}{\prod_{j=1}^{n} b_j} - \sum_{a_j > b_j} \frac{(a_j - b_j)^n}{a_j b_j \prod_{k \neq j} (a_j b_k - b_j a_k)} \right],$$

which coincides with (2.20) in the particular case $J_0 = \emptyset$.

Now, computing the negative value of residues of poles on the right side of $\{\Re(\lambda_1) = \gamma_1\}$ (we need to take the negative value because the path of integration has a negative orientation), yields

$$H(p) = \frac{1}{p^{n+1}} \left[\frac{1}{\prod_{j=1}^{n} a_j} - \sum_{b_j > a_j} \frac{(b_j - a_j)^n}{a_j b_j \prod_{k \neq j} (b_j a_k - a_j b_k)} \right],$$

and after integration w.r.t. p, one also retrieves (2.21) in the particular case $J_0 = \emptyset$.

The associated transform algorithm. Like the direct method algorithm, the above methodology easily extends to an arbitrary number m of nontrivial constraints. The algorithm also consists of m one-dimensional integration steps. Possible pathological multiple poles are handled as in Section 2.4. Clearly, the computational complexity is described by n^m for the same reasons it is in the direct method.

The general case is better illustrated in Example 2.1 of Section 2.4. Setting $e_3 = (1,1,1)$, let $\Omega(ze_3) \subset \mathbb{R}^2$ be the polytope

$$\Omega(ze_3) := \{x \in \mathbb{R}^2_+ | x_1 + x_2 \leq z; -2x_1 + 2x_2 \leq z; 2x_1 - x_2 \leq z\},$$

whose area is $17z^2/48$.

Choose $\gamma_1 = \gamma_2 = 1$, $\gamma_3 = 2$, and $\lambda_3 = p - \lambda_2 - \lambda_1$, so that $\Re(p) = d = 4$ and

$$H(p) = \frac{1}{(2\pi i)^2} \int_{1-i\infty}^{1+i\infty} \int_{1-i\infty}^{1+i\infty} M(\lambda, p)\, d\lambda_1\, d\lambda_2,$$

with

$$M(\lambda, p) = \frac{1}{\lambda_1 \lambda_2 (p - \lambda_1 - \lambda_2)(2p - \lambda_1 - 4\lambda_2)(2\lambda_1 + 3\lambda_2 - p)}.$$

We first integrate w.r.t. λ_1. Only the real parts of the poles $\lambda_1 = 0$ and $\lambda_1 = (p - 3\lambda_2)/2$ are less than 1. Therefore, the residue of the 0-pole yields

$$\frac{1}{2\pi i} \int_{1-i\infty}^{1+i\infty} \frac{1}{\lambda_2 (p - \lambda_2)(2p - 4\lambda_2)(3\lambda_2 - p)}\, d\lambda_2, \tag{2.27}$$

whereas the residue of the $(p - 3\lambda_2)/2$-pole yields

$$\frac{1}{2\pi i} \int_{1-i\infty}^{1+i\infty} \frac{4}{\lambda_2 (p - 3\lambda_2)(p + \lambda_2)(3p - 5\lambda_2)}\, d\lambda_2. \tag{2.28}$$

Applying again Cauchy residue theorem to (2.27) at the pole $\lambda_2 = 0$ (the only one whose real part is less than one) yields $-1/(2p^3)$.

Similarly, applying Cauchy residue theorem to (2.28) at the poles $\lambda_2 = 0$ and $\lambda_2 = -p$ (the only ones whose real part is less than one) yields $(4/3 - 1/8)/p^3$.

We finally obtain $H(p) = (4/3 - 1/8 - 1/2)/p^3 = 17/(24p^3)$, and so $h(z) = 17z^2/48$, the desired result.

Concerning the pathological case of poles on the integration path, we have already said that this issue occurs with probability zero in a set of problems with randomly generated data. Moreover, we have also proved that those poles are *removable singularities* and we only need to add all terms with poles on the integration path in order to get rid of them, with no need to implement the perturbation technique described in Section 2.4.

For example, had we chosen $\gamma_1 = \gamma_2 = \gamma_3 = 1$ (instead of $\gamma_1 = \gamma_2 = 1$ and $\gamma_3 = 2$) in the above example, we would have in both (2.27) and (2.28) a pole on the integration path $\{\Re(\lambda_2) = 1\}$, because $\Re(p) = 3$. But, adding (2.27) and (2.28) together yields

$$H(p) = \frac{1}{2\pi i} \int_{1-i\infty}^{1+i\infty} \frac{5p - 7\lambda_2}{\lambda_2 (p - \lambda_2)(2p - 4\lambda_2)(p + \lambda_2)(3p - 5\lambda_2)}\, d\lambda_2,$$

and, as expected, the integrand has no more poles on the integration path. In addition, only the real parts of poles $\lambda_2 = 0$ and $\lambda_2 = -p$ are less than 1, and their residues give $H(p) = (5/6 - 1/8)/p^3 = 17/(24p^3)$, and so $h(z) = 17z^2/48$.

The above two algorithms easily extend to the case $c \neq 0$ with slight ad hoc modifications, as one only replaces $A'\lambda$ with $A'\lambda - c$ in the denominator of the Laplace transform $\widehat{F}(\cdot, c)$.

2.5 Notes

The computational complexity of volume computation is detailed in Dyer and Frieze [45] and in the nice exposition of Bollobás [24]. For more details on probabilistic approximate methods, the interested reader is referred to Bollobás [24], Dyer, Frieze, and Kannan [47], and Kannan, Lovász, and Simonovits [76]. For a general exposition of quasi-Monte Carlo methods for computing integrals (the volume being a particular one), see, e.g., Niederreiter [116]. Another method to approximate the volume of a convex polytope (with error estimates) is described in Kozlov [84]. It uses an appropriate integral formulation in which the integrand is written as a (truncated) Fourier series.

For an updated review of primal exact methods for volume computation and their computational complexity, see, e.g., Büeler et al. [29] and Gritzmann and Klee [62]. In particular, improved versions of some of the above algorithms are also described in [29], and the software package VINCI developed at ETH in Zürich offers several alternative methods working with the half-space description or the vertex description (or both) of Ω. It is available with free access at
`http://www.lix.polytechnique.fr/Labo/Andreas.Enge/Vinci.html`.

Most of the material in this chapter is from Barvinok [13], Büeler et al. [29], Brion and Vergne [27], and Lasserre and Zeron [95]. The residue method described in this chapter does not require one to work with a vector $c \in \mathbb{R}^n$ having some *regularity* property, as in Lawrence [102] and Brion and Vergne [27]. Despite this regularity property of c being *generic*, the resulting formula is not stable numerically as reported in [29].

Finally, let us mention that the Laplace transform approach developed in this chapter is also applicable to integration of other functions (e.g., Laurent polynomials) over other sets (e.g., ellipsoids). In some cases, one may even get the exact result in closed form or good approximations by truncated series; see, e.g., Lasserre and Zeron [96].

Chapter 3
Comparing the Continuous Problems P and I

3.1 Introduction

In Chapter 2 we saw an algorithm for computing the volume of a convex polytope. With obvious *ad hoc* modifications, it can be used to solve the *integration* problem **I** (see Section 2.4). We qualified this method of dual and even defined a dual integration problem $\mathbf{I}^*$ in (2.13). In this chapter we are now concerned with a comparison between problems **P** and **I** from a *duality* point of view.

So, with $A \in \mathbb{R}^{m\times n}$, $y \in \mathbb{R}^m$, recall that $\Omega(y) \subset \mathbb{R}^n$ is the convex polyhedron defined in (2.2), and consider the linear program (LP) in standard form

$$\mathbf{P}: \; f(y,c) := \max\{\, c'x | Ax = y,\; x \geq 0\} \tag{3.1}$$

with $c \in \mathbb{R}^n$, and its associated *integration* version

$$\mathbf{I}: \; \widehat{f}(y,c) := \int_{\Omega(y)} \mathrm{e}^{c'x}\, d\sigma, \tag{3.2}$$

where $d\sigma$ is the Lebesgue measure on the affine variety $\{x \in \mathbb{R}^n \,|\, Ax = y\}$ that contains the convex polyhedron $\Omega(y)$.

The dual problem $\mathbf{P}^*$

It is well known that the standard duality for (3.1) is obtained from the *Legendre–Fenchel* transform $F(\cdot,c) : \mathbb{R}^m \to \mathbb{R}$ of the value function $y \mapsto f(y,c)$, i.e., here (as $y \mapsto f(\cdot,c)$ is concave)

$$\lambda \mapsto F(\lambda,c) := \inf_{y\in\mathbb{R}^m} \langle \lambda, y\rangle - f(y,c). \tag{3.3}$$

J.B. Lasserre, *Linear and Integer Programming vs Linear Integration and Counting*,
Springer Series in Operations Research and Financial Engineering, DOI 10.1007/978-0-387-09414-4_3,

Developing yields

$$F(\lambda,c)=\begin{cases}0 & \text{if } A'\lambda-c\geq 0,\\ -\infty & \text{otherwise,}\end{cases} \tag{3.4}$$

and applying again the Legendre–Fenchel transform to $F(\cdot,c)$ yields

$$\mathbf{P}^*:\quad \inf_{\lambda\in\mathbb{R}^m}\langle\lambda,y\rangle-F(\lambda,c)=\min_{\lambda\in\mathbb{R}^m}\{y'\lambda : A'\lambda\geq c\}, \tag{3.5}$$

where one recognizes the usual dual $\mathbf{P}^*$ of $\mathbf{P}$.

Hence, the *dual* problem $\mathbf{P}^*$ is itself a linear program which can be solved, e.g., via the simplex algorithm, or any interior points method, exactly in the same manner as for the primal problem $\mathbf{P}$.

In particular, if $\max\mathbf{P}<\infty$ then $\min\mathbf{P}^*>-\infty$ and there is no duality gap, i.e., $\min\mathbf{P}^*=\max\mathbf{P}$. Moreover, the optimal value $\max\mathbf{P}$ is attained at some vertex x^* of the convex polyhedron $\Omega(y)$, whereas the optimal value $\min\mathbf{P}^*$ is attained at some vertex $\lambda^*\in\mathbb{R}^m$ of the convex polyhedron $\{\lambda\in\mathbb{R}^m\,|\,A'\lambda\geq c\}$, and so $c'x^*=y'\lambda^*$.

The dual problem $\mathbf{I}^*$

The analogue for integration of the Legendre–Fenchel transform is the two-sided *Laplace* transform $\widehat{F}(\cdot,c):\mathbb{C}^m\to\mathbb{C}$ of $\widehat{f}(\cdot,c)$, given by

$$\lambda\mapsto\widehat{F}(\lambda,c):=\int_{\mathbb{R}^m}\mathrm{e}^{-\langle\lambda,y\rangle}\widehat{f}(y,c)\,dy=\prod_{j=1}^{n}\frac{1}{(A'\lambda-c)_j}, \tag{3.6}$$

which is well defined provided

$$\Re(A'\lambda-c)>0. \tag{3.7}$$

Indeed, observe that from the definition (3.3) of the Legendre–Fenchel transform, and the monotonicity of the "sup" operator,

$$\mathrm{e}^{-F(\lambda,c)}=\sup_x\mathrm{e}^{-\langle\lambda,x\rangle+f(x)}=\int^{\oplus}\mathrm{e}^{-\langle\lambda,x\rangle}\,\mathrm{e}^{f(x)},$$

where $\int^{\oplus}$ ($\equiv\sup_x$) is the "integration" operator in the $(\max,+)$-algebra (recall that the sup operator is the $\oplus$ (addition) in the $(\max,+)$-algebra, i.e., $\max[a,b]=a\oplus b$. Formally, denoting $\mathscr{F}$ and $\mathscr{L}$ the Legendre–Fenchel and Laplace transforms,

$$\exp-\mathscr{F}[f]=\mathscr{L}[\exp f]\qquad\text{in }(\max,+).$$

Computing $\widehat{f}(y,c)$ reduces to evaluating the *inverse* Laplace transform at the point $y\in\mathbb{R}^m$, which we have called the *dual* integration problem $\mathbf{I}^*$ of (3.2), i.e.,

$$\mathbf{I}^*: \ \widehat{f}(y,c) = \frac{1}{(2i\pi)^m} \int_{\gamma-i\infty}^{\gamma+i\infty} e^{\langle y,\lambda \rangle} \widehat{F}(\lambda,c)\, d\lambda, \tag{3.8}$$

where $\gamma \in \mathbb{R}^m$ is fixed and satisfies $A'\gamma - c > 0$; see Section 2.3.

To pursue the formal analogy between integration and optimization, observe that in the integration process on the path $\Gamma = \{\gamma \pm i\infty\}$, the following are obtained:

- Evaluating $e^{\langle y,\lambda \rangle} \widehat{F}(\lambda,c)$ at a fixed $\lambda \in \Gamma$ is the formal analogue of evaluating $G(\lambda) := \min_{x \geq 0} \{h(x) + \langle \lambda, Ax - y \rangle\}$ in a dual Lagrangian method to minimize h on $\Omega(y)$, whereas summing up over $\lambda \in \Gamma$ is the same as maximizing $G(\lambda)$ over $\mathbb{R}^m$.

3.2 Comparing $\mathbf{P}, \mathbf{P}^*, \mathbf{I}$, and $\mathbf{I}^*$

As already mentioned in Section 2.3, we may indeed call $\mathbf{I}^*$ a dual problem of $\mathbf{I}$ because as $\mathbf{P}^*$, it is defined on the space $\mathbb{C}^m$ of variables $\{\lambda_j\}$ associated with the constraints $Ax = y$, and we also retrieve the standard "ingredients" of $\mathbf{P}^*$, namely, $y'\lambda$ and $A'\lambda - c$.

Moreover, and like $\mathbf{I}$, $\mathbf{I}^*$ is also an *integration* problem but now in $\mathbb{C}^m$ rather than in $\mathbb{R}^n$. Next, while $\{x \in \mathbb{R}^n | Ax = y; x \geq 0\}$ defines the domain of integration of $\mathbf{I}$, $\{\lambda \in \mathbb{C}^m | \Re(A'\lambda - c) > 0\}$ defines the domain of definition of the Laplace transform $\widehat{F}(\lambda,c)$ in $\mathbf{I}^*$ and is precisely the interior of the feasible set of the dual problem $\mathbf{P}^*$ in (3.5).

On the logarithmic barrier function

From (3.4) and (3.6),

$$F(\lambda,c) = \ln \mathrm{I}_{\{A'\lambda - c \geq 0\}}(\lambda),$$

$$\widehat{F}(\lambda,c) = \prod_{j=1}^{n} \frac{1}{(A'\lambda - c)_j} \qquad \text{on } \Re(A'\lambda - c) > 0,$$

where I_A denotes the indicator function of the set A, i.e., $\mathrm{I}_A(x) = 1$ if $x \in A$ and $\mathrm{I}_A(x) = 0$ otherwise. In other words, and denoting the feasible set of $\mathbf{P}^*$ by:

$$\Delta := \{\lambda \in \mathbb{R}^m | \quad A'\lambda - c \geq 0\},$$

- $\exp F(\lambda,c)$ is the indicator function I_Δ of the set Δ.
- $\widehat{F}(\lambda,c)$ is an exponential barrier function of the *same* set Δ.

This should not be a surprise, as a self-concordant barrier function $\phi_K(x)$ of a cone $K \subset \mathbb{R}^n$ is given by the logarithm of Laplace transform $\int_{K^*} e^{-\langle x,s\rangle} ds$ of its dual cone K^* (see, e.g., Güler [63], Truong and Tunçel [130]). Next, notice that

$$\begin{aligned}\widehat{f}(y,rc) &= \frac{1}{(2i\pi)^m} \int_{\gamma_r - i\infty}^{\gamma_r + i\infty} \frac{e^{\langle y,\lambda\rangle}}{\prod_{j=1}^n (A'\lambda - rc)_j} d\lambda \\ &= \frac{r^{m-n}}{(2i\pi)^m} \int_{\gamma - i\infty}^{\gamma + i\infty} \frac{e^{\langle ry,\lambda\rangle}}{\prod_{j=1}^n (A'\lambda - c)_j} d\lambda,\end{aligned}$$

with $\gamma_r = r\gamma$, and we can see that (up to the constant $(m-n)\ln r - m\ln(2i\pi)$) the logarithm of the integrand is nothing less than the well-known *logarithmic barrier function*

$$\lambda \mapsto \phi_\mu(\lambda,y,c) = \mu^{-1}\langle y,\lambda\rangle - \sum_{j=1}^n \ln(A'\lambda - c)_j$$

of the dual problem $\mathbf{P}^*$ (3.5), with barrier parameter $\mu := 1/r$. And, one has

$$\lim_{\mu\to 0} \min_\lambda \phi_\mu(\lambda,y,c) = \min_\lambda \{y'\lambda : A'\lambda \geq c\}.$$

(See, e.g., Den Hertog [39].) Thus, when $r\to\infty$, minimizing the exponential logarithmic barrier function on its domain in $\mathbb{R}^m$ yields the same result as taking its residues.

Recall (see Section A2 of Appendix A) that the Cramer transform $f \mapsto \mathscr{C}(f)$ of a function f is the Legendre–Fenchel transform of the logarithm of its Laplace transform, i.e.,

$$\mathscr{C}(f) := \mathscr{F}[\ln \mathscr{L}(f)].$$

Therefore,

$$\begin{aligned}\min_\lambda \phi_\mu(\lambda,y,c) &= \min_\lambda \langle \frac{y}{\mu},\lambda\rangle - \sum_{j=1}^n \ln(A'\lambda - c)_j \\ &= \min_\lambda \langle y,\frac{\lambda}{\mu}\rangle - \sum_{j=1}^n \ln\left(\mu\left(A'\frac{\lambda}{\mu} - \frac{c}{\mu}\right)_j\right) \\ &= -\ln\mu - \max_\lambda \langle -y,\lambda\rangle + \sum_{j=1}^n \ln(A'\lambda - \frac{c}{\mu})_j \\ &= -\ln\mu - \mathscr{F}\left[\ln\mathscr{L}\left[\widehat{f}\left(\cdot,\frac{c}{\mu}\right)\right]\right](-y), \\ &= -\ln\mu - \mathscr{C}\left[\widehat{f}\left(\cdot,\frac{c}{\mu}\right)\right](-y).\end{aligned} \tag{3.9}$$

That is, up to the constant $\ln \mu$, the function

$$y \mapsto \min_{\lambda} \phi_\mu(\lambda, y, c),$$

which maps y to the minimum of the logarithmic barrier function of $\mathbf{P}^*$ (with right-hand side y), is nothing less than the Cramer transform (with minus sign) of the value function $\widehat{f}(\cdot, c/\mu)$, evaluated at the point $-y$. Similarly, for a minimization problem **P**:

$$\begin{aligned} y \mapsto -\min_{\lambda} \tilde{\phi}_\mu(\lambda, y, c) &:= -\min_{\lambda} \langle \frac{y}{\mu}, \lambda \rangle - \sum_{j=1}^{n} \ln(c - A\lambda)_j \\ &= \ln \mu + \mathscr{C}\left[\widehat{f}\left(\cdot, \frac{-c}{\mu}\right)\right](y). \end{aligned}$$

Hence (3.9) reveals another subtle link between **P** and **I**. Indeed, as $\mu \to 0$, the optimal value $f(y,c)$ of **P** can be approximated either via an integral with $\mu \ln \widehat{f}(y, c/\mu)$ (recall (1.4)); or via optimization with the Cramer transform $\mathscr{C}(\widehat{f}(\cdot, c/\mu))(-y)$ (i.e., minimizing the logarithmic barrier function).

In both cases one uses the function $y \mapsto \widehat{f}(y, c/\mu)$ with parameter c/μ.

Simplex algorithm and Cauchy residues

We have seen in Chapter 2 that one may compute directly $\widehat{f}(y,c)$ by using Cauchy residue techniques. That is, one computes the integral (2.13) by successive one-dimensional complex integrals w.r.t. one variable λ_k at a time (e.g., starting with $\lambda_1, \lambda_2, \ldots$) and by repeated application of Cauchy residue theorem. This is possible because the integrand is a rational fraction, and after application of the Cauchy residue theorem at step k w.r.t. λ_k, the output is still a rational fraction of the remaining variables $\lambda_{k+1}, \ldots, \lambda_m$ (see Section 2.4). The whole procedure is a summation of partial results, each of them corresponding to a (multipole) vector $\widehat{\lambda} \in \mathbb{R}^m$ that annihilates m terms of n products in the denominator of the integrand and corresponds to a vertex of the convex polytope $\Omega(y)$. Hence, Cauchy's residue technique explores all the vertices of $\Omega(y)$. This is nicely summarized in Brion and Vergne's continuous formula (2.15) for $\widehat{f}(y,c)$, with $y \in \gamma$ for some chamber γ. Thus, $\widehat{f}(y,c)$ in (2.15) is a weighted *summation* over the vertices of $\Omega(y)$, whereas $f(y,c)$ is a *maximization* over the vertices (equivalently, also a summation with $\oplus \equiv \max$). That is,

$$\begin{aligned} \widehat{f}(y,c) &= \sum_{x(\sigma):\ \text{vertex of } \Omega(y)} \frac{e^{\langle c, x(\sigma)\rangle}}{\mu(\sigma) \prod_{j \notin \sigma} (-c_j + \pi_\sigma A_j)}, \\ f(y,c) &= \max_{x(\sigma):\ \text{vertex of } \Omega(y)} e^{\langle c, x(\sigma)\rangle}. \end{aligned}$$

Hence, in problem **P**, one computes $f(y,c)$ by searching an optimal vertex $x(\sigma^*)$ where σ^* is an optimal basis of the LP. With each vertex $x(\sigma)$ is associated a dual vector π_σ and reduced costs $c_j - \pi_\sigma A_j$ for all variables x_j not in the basis σ. One then selects the optimal vertex $x(\sigma^*)$ by inspection of the associated reduced costs.

Similarly, in Problem **I**, at each vertex $x(\sigma)$ one computes the same dual vector π_σ and reduced costs $c_j - \pi_\sigma A_j$. One then makes a summation of the residue at each vertex (times the determinant of the basis matrix A_σ).

Next, if c is replaced with rc and $x(\sigma^*)$ denotes the optimal vertex of $\Omega(y)$ at which $c'x$ is maximized, we obtain

$$\widehat{f}(y,rc)^{1/r} = \mathrm{e}^{\langle c,x(\sigma^*)\rangle} \left[\sum_{x(\sigma)} \frac{\mathrm{e}^{r\langle c,x(\sigma)-x(\sigma^*)\rangle}}{r^{n-m}\mu(\sigma)\prod_{k\notin\sigma}(-c_k+\pi_\sigma A_k)} \right]^{1/r},$$

from which it easily follows that

$$\lim_{r\to\infty} \ln \widehat{f}(y,rc)^{1/r} = \langle c,x(\sigma^*)\rangle = \max_{x\in\Omega(y)} \langle c,x\rangle = f(y,c),$$

as already mentioned in (1.4).

Table 3.1 Summary of the parallel between $\mathbf{P},\mathbf{P}^*$ and $\mathbf{I},\mathbf{I}^*$

$\mathbf{P},\mathbf{P}^*$ **Legendre–Fenchel Duality**	$\mathbf{I},\mathbf{I}^*$ **Laplace Duality**
$f(y,c) = \max_{Ax=y;x\geq 0} c'x$	$\widehat{f}(y,c) = \int_{Ax=y;x\geq 0} \mathrm{e}^{c'x}\,ds$
$F(\lambda,c) = \inf_{y\in\mathbb{R}^m} \{\lambda'y - f(y,c)\}$ $= \begin{cases} 0 & \text{if } A'\lambda - c \geq 0 \\ -\infty & \text{otherwise} \end{cases}$ $= \ln \mathrm{I}_{\{\lambda:A'\lambda-c\geq 0\}}$ with domain $A'\lambda - c \geq 0$	$\widehat{F}(\lambda,c) = \int_{\mathbb{R}^m} \mathrm{e}^{-\lambda'y}\widehat{f}(y,c)\,dy$ $= \dfrac{1}{\prod_{k=1}^n (A'\lambda - c)_k}$ with domain $\Re(A'\lambda - c) > 0$
$f(y,c) = \min_{\lambda\in\mathbb{R}^m} \{\lambda'y - F(\lambda,c)\}$ $= \min_{\lambda\in\mathbb{R}^m} \{y'\lambda \mid A'\lambda - c \geq 0\}$	$\widehat{f}(y,c) = \dfrac{1}{(2i\pi)^m}\int_\Gamma \mathrm{e}^{\lambda'y}\widehat{F}(\lambda,c)\,d\lambda$ $= \dfrac{1}{(2i\pi)^m}\int_\Gamma \dfrac{\mathrm{e}^{\lambda'y}}{\prod_{k=1}^n (A'\lambda - c)_k}\,d\lambda$
Simplex algorithm: $\rightarrow$ vertices of $\Omega(y)$ $\rightarrow \max c'x$ over vertices	Cauchy residue: $\rightarrow$ poles of $\widehat{F}(\lambda,c)$ $\rightarrow \sum \mathrm{e}^{c'x}$ over vertices

3.3 Notes

Most of the material of this chapter is from Lasserre [88]. As already noted, problem **P** is the (max,+)-algebra analogue of **I** in the usual algebra $(+,\times)$. In Bacelli et al. [10, §9], *dynamic programming* is also viewed as the (max,+)-algebra analogue of probability in the usual algebra $(+,\times)$. For instance, correspondences are established between quadratic forms and Gaussian densities, inf-convolution and convolution, Fenchel and Fourier transform, Brownian decision processes and Brownian motion, diffusion decision processes and diffusions. See also Burgeth and Weickert [31] for a comparison between linear and morphological systems.

Part II
Linear Counting and Integer Programming

Chapter 4
The Linear Counting Problem $\mathbf{I}_d$

In the second part of the book, we consider the linear counting problem $\mathbf{I}_d$ and establish a comparison with its (max, +)-algebra analogue $\mathbf{P}_d$, as well as with the LP problem $\mathbf{P}$.

4.1 Introduction

The discussion of this chapter is the *discrete* analogue of Chapter 2. We are interested in problem $\mathbf{I}_d$ defined in Chapter 1, that is,

$$\mathbf{I}_d : \widehat{f}_d(y,c) := \left\{ \sum e^{c'x} | x \in \Omega(y) \cap \mathbb{Z}^n \right\}, \tag{4.1}$$

where $\Omega(y) \subset \mathbb{R}^n$ is the convex polyhedron

$$\Omega(y) := \{x \in \mathbb{R}^n | Ax = y, \quad x \geq 0\}, \tag{4.2}$$

for some matrix $A \in \mathbb{Z}^{m \times n}$ and vectors $y \in \mathbb{Z}^m$, $c \in \mathbb{R}^n$. We have already mentioned that $\mathbf{I}_d$ is the *discrete* analogue of $\mathbf{I}$ and that $\mathbf{P}_d$ is the $(\max,+)$-algebra analogue of $\mathbf{I}_d$.

Notice that when $c = 0$, the function $\widehat{f}_d(y,0)$ counts the integral points of the convex polyhedron $\Omega(y)$ then assumed to be compact, i.e., a convex *polytope*. (Its continuous analogue $\widehat{f}(y,0)$ is the volume of $\Omega(y)$.) *Counting* lattice points of a convex polytope is a fundamental problem in computational geometry and operations research, as well, in view of its connection with integer programming. It is a very hard problem even for probabilistic methods. For instance, and in contrast with volume computation, there is still no FPRAS algorithm available for counting feasible solutions of 0–1 knapsack equations (also called *Frobenius equations*). So far, the best known probabilistic algorithm of Dyer et al. [48] has exponential complexity.

The technique of exact methods for counting lattice points in polytopes has received much attention in recent years; see, e.g., the works of Barvinok [13, 14], Barvinok and Pommersheim [15], Beck [17, 18], Beck, Diaz, and Robins [19], Brion [25], Brion and Vergne [26, 27], Kantor and

J.B. Lasserre, *Linear and Integer Programming vs Linear Integration and Counting*,
Springer Series in Operations Research and Financial Engineering, DOI 10.1007/978-0-387-09414-4_4,

Khovanskii [77], and Khovanskii and Pukhlikov [79]. In particular, Barvinok [13, 14] has proposed an algorithm to compute $\widehat{f_d}(y,c)$ with polynomial time computational complexity when the dimension n is fixed. This algorithm requires knowledge of the vertices v of $\Omega(y)$ and uses a special representation (due to Brion [25]) of the *generating function* $g(z) := \sum_{x\in\Omega(y)\cap\mathbb{Z}^n} z^x$. Essential in this remarkable procedure is a *signed* decomposition of certain closed convex cones into *unimodular* cones (see, e.g., Barvinok and Pommersheim [15, Theor. 4.4]). *In fine*, Barvinok's procedure provides a compact description of the rational (generating) function $g(z)$, and getting $\widehat{f_d}(y,c)$ is just evaluating g at the point $z = e^c$. This approach is implemented in the software `LattE` developed at the University of California, Davis [40, 41].

We call this approach *primal* as it works in the primal space $\mathbb{R}^n$ of the x variables, and y is fixed. On the other hand, a *dual*[1] approach is to consider the generating function (or $\mathbb{Z}$-transform) $\widehat{F}_d : \mathbb{C}^m \to \mathbb{C}$ of $\widehat{f_d}(\cdot,c)$, which has a simple and explicit expression in closed form. We call this approach *dual* of Barvinok's primal approach because one works in the space $\mathbb{C}^m$ of variables z associated with the constraints $Ax = y$, and c is fixed. A fundamental result of Brion and Vergne [27] provides a generalized residue formula for the inverse $\mathbb{Z}$-transform of $\widehat{F}_d$, which yields a nice and elegant explicit expression for $\widehat{f_d}(y,c)$, the discrete analogue of (2.15). But in general, Brion and Vergne's discrete formula [27] may be very hard to evaluate numerically. Recall that it is even the case for its simpler continuous analogue (2.15); see the comment after the Lawrence formula (2.5). As mentioned in Baldoni-Silva and Vergne [11], it requires many steps (in particular, one has to build up *chambers* (maximal cones) in a subdivision of a closed convex cone into polyhedral cones [27]). One also has to handle roots of unity in Fourier–Dedekind sums, a nontrivial task. Beck [17] and Beck, Diaz, and Robins [19] also provide a complete analysis based on residue techniques for the case of a tetrahedron ($m = 1$) and also mention the possibility of evaluating $\widehat{f_d}(y,c)$ for general polytopes by means of residues. The case where A is *unimodular* simplifies and is exploited in Baldoni-Silva and Vergne [11] for particular cases like flow polytopes, as well as in Baldoni et al. [12].

Inverting the generating function is the discrete analogue of inverting the Laplace transform as we did for the continuous case in Chapter 2, and both reduce to evaluating a complex integral. However, and in contrast with the continuous case, in the discrete case this approach of computing a complex integral by means of residues is fastidious and numerically involved because the generating function $\widehat{F}_d$ has many more poles (in particular complex) than in the continuous case.

In this chapter, we also take a dual approach. We consider problem $\mathbf{I}_d$ as that of evaluating the value function $\widehat{f}(y,c)$ at some particular $y \in \mathbb{R}^m$, and to do so, we compute the inverse $\mathbb{Z}$-transform of its generating function $\widehat{F}_d(z,c)$ at the point $y \in \mathbb{Z}^m$; inversion of the $\mathbb{Z}$-transform is what we call the *dual problem* $\mathbf{I}_d^*$. To avoid computing residues, we provide a decomposition of the rational function $\widehat{F}_d$ into simpler rational functions whose inversion is easy.

4.2 A primal approach: Barvinok's counting algorithm

We start with some important material related to the algebra of convex polyhedra of $\mathbb{R}^n$ and its application to derive nice, explicit expressions for *generating functions* associated with them.

[1] Again, duality here should not be confused with the duality between the *vertex* and *half-space* descriptions of a convex polytope $\Omega \subset \mathbb{R}^n$.

For a convex rational polyhedron $\mathscr{A} \subset \mathbb{R}^n$ denote by $[\mathscr{A}]$ its indicator function, i.e., $[\mathscr{A}](x) = 1$ if $x \in \mathscr{A}$ and $[\mathscr{A}](x) = 0$ otherwise. Similarly, denote by $\mathrm{co}(\mathscr{A}, v)$ its *supporting* (or *tangent*) cone at the vertex $v \in \mathscr{A}$, that is,

$$\mathrm{co}(\mathscr{A}, v) = \{x \in \mathbb{R}^n | \varepsilon x + (1-\varepsilon)v \in \mathscr{A} \text{ for all sufficiently small } \varepsilon > 0\}.$$

Next, denote by $\mathscr{P}(\mathbb{Q}^n)$ the algebra generated by the indicators $[\mathscr{A}]$ of rational convex polyhedra $\mathscr{A} \subset \mathbb{R}^n$, and denote by $\mathscr{P}$ the vector space spanned by the indicators of rational convex polyhedra that contain *lines*. Then one has the following important result:

For a convex rational polyhedron $\mathscr{A} \subset \mathbb{R}^n$,

$$[\mathscr{A}] = h + \sum_{v:\ \text{vertex of } \mathscr{A}} [\mathrm{co}(\mathscr{A}, v)] \tag{4.3}$$

for some $h \in \mathscr{P}$.

Example 4.1. Let $\mathscr{A} := [0,1] \subset \mathbb{R}$, so that $\mathrm{co}(\mathscr{A}, \{1\}) = (-\infty, 1]$ and $\mathrm{co}(\mathscr{A}, \{0\}) = [0, +\infty)$. Then obviously

$$[A] = [(-\infty, 1]] + [[0, +\infty)] - [\mathbb{R}].$$

For every rational convex polyhedron $\mathscr{A} \subset \mathbb{R}^n$, let $h(\mathscr{A}, \cdot) : \mathbb{C}^n \to \mathbb{C}$ be the mapping

$$z \mapsto h(\mathscr{A}, z) := \{\textstyle\sum z^x | x \in \mathscr{A} \cap \mathbb{Z}^n\}. \tag{4.4}$$

With V being a vector space, a linear transformation $\mathscr{P}(\mathbb{Q}^n) \to V$ is called a *valuation*. Another important result of Lawrence and Khovanski and Pukhlikov states the following:

Proposition 4.1. *Let $\mathbb{C}(\mathbf{x})$ be the ring of rational functions in the variables $\mathbf{x} = (x_1, \ldots, x_n)$. There exists a valuation $\mathscr{F} : \mathscr{P}(\mathbb{Q}^n) \to \mathbb{C}(\mathbf{x})$ such that*

(i) If $\mathscr{A} \subset \mathbb{R}^n$ is a rational convex polyhedron without lines then $\mathscr{F}(\mathscr{A}) = h(\mathscr{A}, \cdot)$.
(ii) If $\mathscr{A} \subset \mathbb{R}^n$ is a rational convex polyhedron containing a line then $\mathscr{F}(\mathscr{A}) = 0$.

Formula (4.3) and Proposition 4.1 are central to establishing the next two results. Brion [25] proved that for a nonempty rational convex polyhedron $\mathscr{A} \subset \mathbb{R}^n$ with no line

$$h(\mathscr{A}, z) = \sum_{v:\ \text{vertex of } \mathscr{A}} h(\mathrm{co}(\mathscr{A}, v), z). \tag{4.5}$$

See also Barvinok and Pommersheim [15, Theor. 3.5, p. 12]. Hidden in this formula is the fact that, in general, there is no $z \in \mathbb{C}^n$ such that the sum in (4.4) defining $h(\mathrm{co}(\mathscr{A}, v), z)$ converges for each v. In other words, the sum (4.5) is in fact formal, and yet provides $h(\mathscr{A}, z)$. Referring back to Example 4.1, we obtain

$$\begin{aligned} h([0,1],z) = 1+z &= h((-\infty,1],z)+h([0,+\infty),z) \\ &= \frac{z^2}{z-1}+\frac{1}{1-z} = 1+z \end{aligned}$$

despite the fact that the first series converges for $|z|>1$ while the second series converges for $|z|<1$.

Barvinok's counting algorithm. Using Brion's formula (4.5), Barvinok showed that for a nonempty rational convex polytope $\mathscr{A} \subset \mathbb{R}^n$ with no line, $h(\mathscr{A},z)$ has the compact form:

$$h(\mathscr{A},z) = \sum_{i\in I} \varepsilon_i \frac{z^{a_i}}{\prod_{k=1}^n (1-z^{b_{ik}})}, \tag{4.6}$$

where I is a certain index set, and for all $i \in I$, $\varepsilon_i \in \{-1,+1\}$, $\{a_i, \{b_{ik}\}_{k=1}^n\} \subset \mathbb{Z}^n$. Each $i \in I$ is associated with a *unimodular cone* in a signed decomposition of the tangent cones of $\mathscr{A}$ (at its vertices) into unimodular cones.

In Barvinok's algorithm, the number $|I|$ of unimodular cones in that decomposition is $\mathscr{L}^{O(n)}$ where $\mathscr{L}$ is the input size of $\mathscr{A}$, and the overall computational complexity to obtain the coefficients $\{a_i, b_{ik}\}$ in (4.6) is also $\mathscr{L}^{O(n)}$. Crucial for the latter property is the *signed* decomposition (triangulation alone into unimodular cones does not guarantee this polynomial time complexity). For more details, the interested reader is referred to Barvinok [13] and Barvinok and Pommersheim [15].

So, if we take $\mathscr{A} \subset \mathbb{R}^n$ to be the convex polyhedron $\Omega(y) \subset \mathbb{R}^n$ defined in (4.2) for some matrix $A \in \mathbb{Z}^{m\times n}$ and vector $y \in \mathbb{Z}^m$, then evaluating (4.6) at the point $z := \mathrm{e}^c \in \mathbb{R}^n$ yields

$$h(\mathscr{A},\mathrm{e}^c) = \widehat{f_d}(y,c).$$

That is, Barvinok's algorithm permits us to evaluate $\widehat{f_d}(y,c)$ and so to solve problem $\mathbf{I}_d$ defined in (4.1). Therefore, when the dimension n is fixed and $z = \mathrm{e}^c = \{\mathrm{e}^{c_j}\}$ is given, one may solve $\mathbf{I}_d$ in time polynomial in the input size of $\Omega(y)$.

In addition, when the dimension n is fixed, Barvinok and Woods [16] have provided a polynomial time algorithm, which given a convex rational polytope $\mathscr{A} \subset \mathbb{R}^n$, and a linear transformation $T : \mathbb{R}^n \to \mathbb{R}^k$ with $T(\mathbb{Z}^n) \subset \mathbb{Z}^k$, computes the generating function $h(T(\mathscr{A} \cap \mathbb{Z}^n),z)$ in the form (4.6).

As an important consequence, Barvinok and Woods were able to develop an efficient *calculus* on generating functions. For instance, given two convex polytopes $\mathscr{A}_1, \mathscr{A}_2 \subset \mathbb{R}^n$ and their generating functions $h(\mathscr{A}_1,z), h(\mathscr{A}_2,z)$ in the form (4.6), one may obtain the generating function of their intersection $h(\mathscr{A}_1 \cap \mathscr{A}_2,z)$ in the form (4.6), in polynomial time in the input size.

As already mentioned in the introduction, we call this approach *primal* because it works in the primal space $\mathbb{R}^n$ of variables x in the description of the polyhedron $\Omega(y)$. The right hand side $y \in \mathbb{Z}^m$ is *fixed* while $z \in \mathbb{C}^n$ is the variable of interest in the generating function $h(\mathscr{A},\cdot)$. A dual approach works in the space $\mathbb{R}^m$ (or $\mathbb{C}^m$) of the (dual) variables associated with the m linear constraints that define $\Omega(y)$.

4.3 A dual approach

Recall that $\mathbb{Z}_+ = \mathbb{N} = \{0,1,2,\dots\}$ denotes the set of natural numbers, and given $z \in \mathbb{C}^m$ and $u \in \mathbb{Z}^m$, the notation z^u and $\ln(z)$ stands for

$$\begin{aligned} z^u &:= z_1^{u_1} z_2^{u_2} \cdots z_m^{u_m}, \\ \ln(z) &:= [\ln(z_1), \ln(z_2), \dots, \ln(z_m)]. \end{aligned}$$

Similarly, for a matrix $A \in \mathbb{Z}^{m \times n}$ and a vector $z \in \mathbb{C}^m$,

$$\begin{aligned} z^{A_j} &:= \left(z_1^{A_{1j}} z_2^{A_{2j}} \cdots z_m^{A_{mj}}\right), \\ z^A &:= (z^{A_1}, z^{A_2}, \dots, z^{A_m}). \end{aligned}$$

The $\mathbb{Z}$-transform approach

Consider the convex polyhedron $\Omega(y)$ defined in (4.2) with $y \in \mathbb{Z}^m$ and $A \in \mathbb{Z}^{m \times n}$. We want to compute

$$y \mapsto \widehat{f}_d(y,c) := \sum_{x \in \mathbb{Z}^n \cap \Omega(y)} \mathrm{e}^{c'x} \tag{4.7}$$

for some given vector $c \in \mathbb{R}^n$.

If $c = 0$ then $\widehat{f}_d(y,0) = \operatorname{card} \Omega(y)$, the number of integral points in $\Omega(y)$. Of course, computing the number of points $x \in \mathbb{N}^n$ of the convex polytope

$$\Omega_1(y) = \{x \in \mathbb{R}^n_+ \,|\, A_1 x \le y\} \tag{4.8}$$

for some $A_1 \in \mathbb{Z}^{m \times n}$ reduces to computing the cardinality of $\mathbb{N}^n \cap \Omega(y)$ with $\Omega(y)$ as in (4.2) but with $A := [A_1 \,|\, I]$ ($I \in \mathbb{N}^{m \times m}$ being the identity matrix).

An appropriate tool to analyze the function $\widehat{f}_d(\cdot,c)$ is its two-sided $\mathbb{Z}$-*transform* or *generating function*, $\widehat{F}_d(\cdot,c) : \mathbb{C}^m \to \mathbb{C}$, of $y \mapsto \widehat{f}_d(y,c)$, defined by the Laurent series:

$$z \mapsto \widehat{F}_d(z,c) := \sum_{y \in \mathbb{Z}^m} \widehat{f}_d(y,c)\, z^{-y}. \tag{4.9}$$

Let $e_m = (1,1,\dots)$ be the vector of ones in $\mathbb{R}^m$.

Theorem 4.1. *Let $\widehat{f}_d$ and $\widehat{F}_d$ be as in (4.7) and (4.9), respectively, and assume that (2.8) holds. Then*

$$\widehat{F}_d(z,c) = \prod_{j=1}^{n} \frac{1}{(1 - \mathrm{e}^{c_j} z^{-A_j})} \tag{4.10}$$

on the domain

$$|z|^A > \mathrm{e}^c. \tag{4.11}$$

Moreover, for every $y \in \mathbb{Z}^m$,

$$\widehat{f}_d(y,c) = \frac{1}{(2\pi i)^m} \int_{|z_1|=w_1} \cdots \int_{|z_m|=w_m} \widehat{F}_d(z,c) z^{y-e_m} \, dz, \tag{4.12}$$

where $w \in \mathbb{R}^m_+$ satisfies $w^A > c$.

Proof. Apply the definition (4.9) of $\widehat{F}_d$ to obtain

$$\widehat{F}_d(z,c) = \sum_{y \in \mathbb{Z}^m} z^{-y} \left[\sum_{Ax=y, x \in \mathbb{N}^n} \mathrm{e}^{c'x} \right] = \sum_{x \in \mathbb{N}^n} \mathrm{e}^{c'x} z^{-(Ax)}.$$

On the other hand, notice that

$$\mathrm{e}^{c'x} z_1^{-(Ax)_1} z_2^{-(Ax)_2} \cdots z_m^{-(Ax)_m} = \prod_{j=1}^{m} \left(\mathrm{e}^{c_j} z_1^{-A_{1j}} z_2^{-A_{2j}} \cdots z_m^{-A_{mj}} \right)^{x_j}$$

Hence, the conditions $|z_1^{A_{1j}} z_2^{A_{2j}} \ldots z_m^{A_{mj}}| > \mathrm{e}^{c_j}$ for $j = 1,2,\ldots n$ (equivalently, $|z|^A > \mathrm{e}^c$) yield

$$\widehat{F}_d(z,c) = \prod_{j=1}^{n} \sum_{x_j=0}^{\infty} \left(\mathrm{e}^{c_j} z_1^{-A_{1j}} z_2^{-A_{2j}} \cdots z_m^{-A_{mj}} \right)^{x_j}$$

$$= \prod_{j=1}^{n} \frac{1}{(1 - \mathrm{e}^{c_j} z_1^{-A_{1j}} z_2^{-A_{2j}} \cdots z_m^{-A_{mj}})},$$

which is (4.10). Finally, equation (4.12) is obtained by analyzing the integral $\int_{|z|=r} z^w dz$ with $r > 0$. This integral is equal to $2\pi i$ only if $w = -1$, whereas if w is any integer different than -1, then the integral is equal to zero. It remains to show that the domain $\{v \in \mathbb{R}^m_+ | v^A > \mathrm{e}^c\}$ (equivalently, $\{v \in \mathbb{R}^m_+ | A' \ln v > c\}$) is not empty. But as (2.8) holds, this follows directly from Remark 2.1. □

The dual problem $\mathbf{I}_d^*$

We define the dual problem $\mathbf{I}_d^*$ to be that of *inverting* $\widehat{F}_d$, i.e., evaluating the complex integral

$$\mathbf{I}_d^* : \widehat{f}_d(y,c) = \frac{1}{(2\pi i)^m} \int_{|z_1|=w_1} \cdots \int_{|z_m|=w_m} \widehat{F}_d(z,c) z^{y-e_m} dz, \tag{4.13}$$

where $w \in \mathbb{R}^m_+$ satisfies $w^A > c$.

Exactly like the inversion problem $\mathbf{I}^*$ is a dual of $\mathbf{I}$, the inversion problem $\mathbf{I}_d^*$ is a dual problem of $\mathbf{I}_d$. In evaluating the complex integral (4.13) one also works in the space $\mathbb{C}^m$ of variables z associated with the constraints $Ax = y$ in the definition of $\Omega(y)$. It is the discrete analogue of $\mathbf{I}^*$ in Section 2.3; see the next chapter for a more detailed comparison between $\mathbf{I}^*$ and $\mathbf{I}_d^*$. As for the continuous case, Brion and Vergne have provided a nice generalized residue formula for inverting $\widehat{F}_d(z,c)$, which in turn yields an exact formula in closed form for $\widehat{f}_d(y,c)$.

Brion and Vergne's discrete formula

Brion and Vergne [27] consider the *generating function* $H\colon \mathbb{C}^m \to \mathbb{C}$ defined by

$$\lambda \mapsto H(\lambda,c) := \sum_{y\in\mathbb{Z}^m} \widehat{f}_d(y,c)\mathrm{e}^{-\langle\lambda,y\rangle},$$

which, after the change of variable $z_i = \mathrm{e}^{\lambda_i}$ for all $i = 1,\ldots,m$, reduces to $\widehat{F}_d(z,c)$ in (4.10).

Let $\Lambda \subset \mathbb{Z}^m$ be the sublattice $A(\mathbb{Z}^n)$ of the lattice $\mathbb{Z}^m$. With same notation used in Section 2.3, let $c \in \mathbb{R}^n$ be regular with $-c$ in the interior of $(\mathbb{R}^n_+ \cap V)^*$, and let γ be a chamber. For a basis σ, let $\mu(\sigma)$ denote the volume of the convex polytope $\{\sum_{j\in\sigma} t_j A_j, 0 \le t_j \le 1, \forall j \in \sigma\}$, normalized so that $\mathrm{vol}(\mathbb{R}^m/\Lambda) = 1$. For $\sigma \in \mathscr{B}(\Delta,\gamma)$, let $x(\sigma) \in \mathbb{R}^n_+$ be such that $y = \sum_{j\in\sigma} x_j(\sigma)A_j$, and $x_j(\sigma) = 0$, $j \notin \sigma$.

Next, let $G(\sigma) := (\oplus_{j\in\sigma}\mathbb{Z}A_j)^*/\Lambda^*$ (where * denotes the dual lattice); that is, $G(\sigma)$ is a finite abelian group of order $\mu(\sigma)$, and with (finitely many) characters $\mathrm{e}^{2i\pi y}$ for all $y \in \Lambda$; in particular, writing $A_k = \sum_{j\in\sigma} u_{jk}A_j$ for all $k \notin \sigma$,

$$\mathrm{e}^{2i\pi A_k}(g) = \mathrm{e}^{2i\pi \sum_{j\in\sigma} u_{jk} g_j}, \qquad k \notin \sigma.$$

To compare the continuous and discrete Brion and Vergne's formula, let π_σ be as in (2.15), i.e., the dual vector associated with a feasible basis σ of the linear program $\mathbf{P}$. (Observe that π_σ is now rational.) Then for all $y \in \Lambda \cap \overline{\gamma}$

$$\widehat{f}_d(y,c) = \sum_{\sigma\in\mathscr{B}(\Delta,\gamma)} \frac{\mathrm{e}^{c'x(\sigma)}}{\mu(\sigma)} U_\sigma(y,c), \tag{4.14}$$

with

$$U_\sigma(y,c) = \sum_{g\in G(\sigma)} \frac{\mathrm{e}^{2i\pi y}(g)}{\prod_{k\notin\sigma}(1 - \mathrm{e}^{-2i\pi A_k}(g)\mathrm{e}^{c_k - \pi_\sigma A_k})}. \tag{4.15}$$

Due to the occurrence of *complex poles* in $\widehat{F}(z,c)$, the term $U_\sigma(y,c)$ in (4.14) and (4.15) is the *periodic* analogue of $\prod_{k\notin\sigma}(c_k - \pi_\sigma A_k)^{-1}$ in (2.15). Again, note the importance of the reduced cost $(c_j - \pi_\sigma A_j)$, $j \notin \sigma$, in both formulas.

Both formulas (2.14) and (4.14) state a *weighted* summation of $e^{c'x(\sigma)}$ over all bases $\sigma \in \mathscr{B}(\Delta,\gamma)$ (or over all vertices of $\Omega(y)$ if $y \in \gamma$). The only difference is the weight. Although both weights are stated in terms of the reduced costs $(c_j - \pi_\sigma A_j)$ of the variables j that are not in the basis σ, the weight in the discrete case is composed of $\mu(\sigma)$ complex terms as opposed to it having only one real term in the continuous case.

In the next section, we will see that this difference explains the difficulties in computing $\widehat{f}_d$ by evaluating the complex integral in (4.13) via Cauchy residue techniques. And not surprisingly, this also makes a direct numerical evaluation of $\widehat{f}_d(y,c)$ via (4.14)–(4.15) very difficult and not practical in general.

4.4 Inversion of the $\mathbb{Z}$-transform by residues

In principle, Theorem 4.1 allows us to compute $\widehat{f}_d(y,c)$ for $y \in \mathbb{Z}^m$ via (4.12), that is, by computing the inverse $\mathbb{Z}$-transform of $\widehat{F}_d(z,c)$ at the point y. Moreover, we can directly calculate (4.12) by repeated applications of Cauchy residue theorem because $\widehat{F}_d(z,c)$ is a rational function with only *finitely many* poles (with respect to one variable at a time). We call this technique the *direct* $\mathbb{Z}$ *inverse*. On the other hand, we can also slightly simplify the inverse problem and invert what is called the *associated* $\mathbb{Z}$*-transform*, which yields some advantages when compared to the direct inversion. In particular, it permits us to compute the *Ehrhart polynomial*, an important mathematical object.

The associated $\mathbb{Z}$*-transform*

With no loss of generality, we assume that $y \in \mathbb{Z}^m$ is such that $y_1 \neq 0$. We may also suppose that each y_i is a multiple of y_1 (taking 0 to be multiple of any other integer). Otherwise, we just need to multiply each constraint $(Ax)_i = y_i$ by $y_1 \neq 0$ when $i = 2,3,\ldots,m$, so that the new matrix A and vector y still have entries in $\mathbb{Z}$.

Hence, there exists a vector $D \in \mathbb{Z}^m$ with first entry $D_1 = 1$ and such that $y = Dy_1$. Notice that D may have entries equal to zero or even negative, but not its first entry. The inversion problem is thus reduced to evaluating, at the point $t := y_1$, the function $g : \mathbb{Z} \to \mathbb{N}$ defined by

$$g(t) := \widehat{f}_d(Dt,c) = \frac{1}{(2\pi i)^m} \int_{|z_m|=w_m} \cdots \int_{|z_1|=w_1} \widehat{F}_d(z,c) z^{Dt-e_m} dz, \tag{4.16}$$

where $\widehat{F}_d$ is given in (4.10), $e_m := (1,1,\ldots)$ is the vector of ones in $\mathbb{R}^m$, and the real (fixed) vector $w \in \mathbb{R}^m_+$ satisfies $A'\ln(w) > c$. The following technique permits us to calculate (4.16).

Consider the following simple change of variables. Let $p = z^D$ and $d = w^D$ in (4.16), so that $z_1 = p\prod_{j=2}^m z_j^{-D_j}$ and

$$g(t) = \frac{1}{(2\pi i)^m} \int_{|z_m|=w_m} \cdots \int_{|z_2|=w_2} \left[\int_{|p|=d} p^{t-1} \widehat{\mathscr{F}}\, dp \right] dz_2 \ldots dz_m,$$

where

$$\widehat{\mathscr{F}}(z_2,\ldots,z_m,p) = \widehat{F}_d\left(p\prod_{j=2}^{m} z_j^{-D_j}, z_2,\ldots,z_m,c\right)\prod_{j=2}^{m} z_j^{-D_j}. \tag{4.17}$$

We can rewrite $g(t)$ as

$$g(t) = \frac{1}{2\pi i}\int_{|p|=d} p^{t-1}\mathscr{G}(p)dp, \tag{4.18}$$

with

$$\mathscr{G}(p) := \frac{1}{(2\pi i)^{m-1}}\int_{|z_2|=w_2}\cdots\int_{|z_m|=w_m}\widehat{\mathscr{F}}\, dz_2\ldots dz_m, \tag{4.19}$$

and $\mathscr{G}$ is called the *associated $\mathbb{Z}$-transform* of $\widehat{f}_d$ with respect to D.

Recall that $\widehat{F}_d(z,c)$ is well-defined on the domain (4.11), and so the domain of definition of $\widehat{\mathscr{F}}$ is given by

$$\left(|p|\prod_{j=2}^{m}|z_j|^{-D_j}, |z_2|,\ldots,|z_m|\right) \in \{\beta\in\mathbb{R}_+^n \,|\, A'\ln\beta > c\}.$$

The Ehrhart polynomial

Consider the convex polytope $\Omega_1(y)$ defined in (4.8) and the *dilated* polytope $t\Omega_1(y) := \{tx \,|\, x\in\Omega_1(y)\}$. Then, with $A := [A_1\,|\,I]$ and $c := 0$, consider the function $t\mapsto g(t)$ defined in (4.16).

When $\Omega_1(y)$ is an *integer polytope* (an integer polytope has all its vertices in $\mathbb{Z}^m$) then g is a *polynomial* in t, that is,

$$g(t) = a_n t^n + a_{n-1}t^{n-1} + \cdots + a_0, \tag{4.20}$$

with $a_0 = 1$ and $a_n = \text{volume}(\Omega_1(y))$. It is called the *Ehrhart polynomial* of $\Omega_1(y)$ (see, e.g., Barvinok and Pommersheim [15], Ehrhart [52]).

On the other hand, if $\Omega_1(y)$ is not an integer polytope (i.e., $\Omega_1(y)$ has rational vertices), then the function $t\mapsto g(t)$ is a *quasipolynomial*, that is, a polynomial in the variable t like in (4.20) but now with coefficients $\{a_i(t)\}$ that are *periodic* functions of t; for instance, see (4.21) in Example 4.2 below. To compute $g(t)$ one may proceed as follows.

(1) Compute $\mathscr{G}(p)$ in (4.19), in $m-1$ steps, where each step k is a one-dimensional integration w.r.t. z_k, $k = 2,\ldots,m$.
(2) Compute $g(t)$ in (4.18), a one-dimensional integration w.r.t. p.

At each of these steps, the one-dimensional complex integrals are evaluated by Cauchy residue theorem. As already mentioned, and due to the presence of complex poles, the inversion is much more involved than that of $\widehat{F}$ in Chapter 2, as illustrated in the following simple example.

Example 4.2. Consider the following convex rational polytope with three ($m = 3$) nontrivial constraints.

$$\Omega_1(te_3) := \{x\in\mathbb{R}_+^2 \,|\, x_1 + x_2 \le t, -2x_1 + 2x_2 \le t, \text{ and } 2x_1 - x_2 \le t\}.$$

With

$$c = 0, \quad A := \begin{bmatrix} 1 & 1\,1\,0\,0 \\ -2 & 2\,0\,1\,0 \\ 2 & -1\,0\,0\,1 \end{bmatrix},$$

and by Theorem 4.1, we have to calculate the inverse $\mathbb{Z}$-transform of

$$\widehat{F_d}(z,c) = \frac{z_1 z_2 z_3}{(z_1-1)(z_2-1)(z_3-1)(1-z_1^{-1}z_2^2 z_3^{-2})(1-z_1^{-1}z_2^{-2}z_3)},$$

where

$$\begin{cases} |z_j| > 1 & \text{for } j = 1,2,3, \\ |z_1 z_2^{-2} z_3^2| > 1, \\ |z_1 z_2^2 z_3^{-1}| > 1. \end{cases}$$

We wish to work with rational functions whose denominator degree is the smallest possible. To do so, we fix $z_1 = p/(z_2 z_3)$ and divide by $z_2 z_3$ (see 4.17) because z_1 has the exponents with smallest absolute value. Therefore,

$$\widehat{\mathscr{F}}(z_2,z_3,p) = \frac{z_3 p^2}{(z_2^{-1}p - z_3)(z_2-1)(z_3-1)(z_3 - z_2^3 p^{-1})(z_2 p - z_3^2)},$$

where

$$\begin{cases} |p| > |z_2 z_3|, \\ |z_3 p| > |z_2^3| > 1, \\ |z_2 p| > |z_3^2| > 1. \end{cases}$$

Notice that $z_2^* = z_3^* = 2$ and $p^* = 5$ is a solution of the previous system of inequalities. We next compute (4.18) and (4.19) by fixing $w_2 = z_2^*$, $w_3 = z_3^*$, and $d = p^*$. Let us integrate $\widehat{\mathscr{F}}$ along the circle $|z_3| = z_3^*$ with a positive orientation. Observe that (taking $p := p^*$ and $z_2 := z_2^*$ constant) $\widehat{\mathscr{F}}$ has two poles located on the circle of radius $|z_2 p|^{1/2} > z_3^*$, and three poles located on the circles of radii $1 < z_3^*$, $|z_2^{-1}p| > z_3^*$, and $|z_2^3 p^{-1}| < z_3^*$. Hence, we can consider poles inside the circle $|z_3| = z_3^*$ in order to avoid considering the pole $z_3 := (z_2 p)^{1/2}$ with fractional exponent. This yields

$$I_1(p,z_2) = \frac{z_2 p^2}{(p-z_2)(z_2-1)(p-z_2^3)(z_2-p^{-1})} + \frac{z_2^3 p^5}{(p^2-z_2^4)(z_2-1)(z_2^3-p)(p^3-z_2^5)}.$$

Next, we integrate I_1 along the circle $|z_2| = z_2^*$. Taking $p := p^*$ as a constant, the first term of I_1 has poles on circles of radii $|p| > z_2^*$, $1 < z_2^*$, $|p|^{1/3} < z_2^*$, and $|p|^{-1} < z_2^*$. We consider poles outside the circle $|z_2| = z_2^*$ in order to avoid considering the pole $z_2 := p^{1/3}$, and we obtain

$$\frac{-p^3}{(p-1)(p^2-1)^2}.$$

The second term of I_1 has poles on circles of radii $|p|^{1/2} > z_2^*$, $1 < z_2^*$, $|p|^{1/3} < z_2^*$, and $|p|^{3/5} > z_2^*$. Notice that we have poles with fractional exponents inside and outside the integration path $|z_2| = z_2^*$. Expanding the second term of I_1 into simple fractions yields

$$\frac{p^5}{(z_2-1)(p^2-1)(1-p)(p^3-1)} + \frac{(2p^3+2p^4+3p^5+p^6+p^7)z_2^2+\alpha_1(p)z_2+\alpha_0(p)}{(z_2^3-p)(p-1)^3(p+1)^2(p^2+1)} + \frac{Q_1(z,p)}{(z_2^4-p^2)} + \frac{Q_2(z,p)}{(z_2^5-p^3)}.$$

Obviously, we only calculate terms with denominator (z_2-1) or (z_2^3-p). As a result the associated $\mathbb{Z}$-transform $\mathscr{G}(p)$ is given by

$$\mathscr{G}(p) = \frac{-p^3}{(p-1)^3(p+1)^2} - \frac{p^5}{(p-1)^2(p+1)(p^3-1)} + \frac{2p^3+2p^4+3p^5+p^6+p^7}{(p-1)^3(p+1)^2(p^2+1)}.$$

Finally, it remains to integrate $\mathscr{G}(p)p^{t-1}$ along the circle $|p| = p^*$. Observe that $\mathscr{G}(p)p^{t-1}$ has poles when p is equal to $1, -1, i = \sqrt{-1}, -i, \sigma = e^{2\pi i/3}$, and $\bar{\sigma}$. Hence, for $t \in \mathbb{Z}_+$, we finally obtain

$$g(t) = \frac{17t^2}{48} + \frac{41t}{48} + \frac{139}{288} + \frac{t(-1)^t}{16} + \frac{9(-1)^t}{32} + \frac{\sqrt{2}}{8}\cos\left(\frac{\pi t}{2}+\frac{\pi}{4}\right) + \frac{2}{9}\cos\left(\frac{2\pi t}{3}+\frac{\pi}{3}\right). \tag{4.21}$$

Notice that we have used the two identities

$$\sqrt{2}\cos\left(\frac{\pi t}{2}+\frac{\pi}{4}\right) = \frac{(i-1)i^t}{2i} + \frac{(i+1)(-i)^t}{2i},$$
$$2\cos\left(\frac{2\pi t}{3}+\frac{\pi}{3}\right) = \frac{(\sigma-1)\sigma^t}{\sigma-\bar{\sigma}} + \frac{(\bar{\sigma}-1)\bar{\sigma}^t}{\bar{\sigma}-\sigma}.$$

It is now easy to count the number $f(te_3) = g(t)$ of points $x \in \mathbb{N}^2$ of $\Omega(te_3)$:

$$f(0) = \frac{139}{288} + \frac{9}{32} + \frac{1}{8} + \frac{1}{9} = 1,$$
$$f(e_3) = \frac{17}{48} + \frac{41}{48} + \frac{139}{288} - \frac{1}{16} - \frac{9}{32} - \frac{1}{8} - \frac{2}{9} = 1,$$
$$f(2e_3) = \frac{68}{48} + \frac{82}{48} + \frac{139}{288} + \frac{2}{16} + \frac{9}{32} - \frac{1}{8} + \frac{1}{9} = 4.$$

Observe that in (4.21) the function $t \mapsto g(t)$ is a *quasipolynomial* because the linear and constant terms have periodic coefficients. However, with $t = 12$, one may check that the rational polytope $\Omega_1(te_3)$ is in fact an *integer* polytope, that is, all its vertices are in $\mathbb{Z}^n$. Therefore, $\Omega_1(12te_3)$ is the t-dilated polytope of $\Omega_1(12e_3)$, and $\tilde{g}(t) := g(12t)$ (with $g(t)$ as in (4.21)) is thus the Ehrhart polynomial of $\Omega_1(12e_3)$, which reads

$$\tilde{g}(t) = 51t^2 + 11t + 1,$$

and indeed, 51 is the volume of $\Omega_1(12e_3)$ and the constant term is 1, as it should be.

4.5 An algebraic method

As seen on the previous simple illustrative example, inversion of $\widehat{F}_d$ by the Cauchy residue theorem can rapidly become tedious. In this section we propose an alternative method, one which avoids complex integration and is purely algebraic. It is particularly attractive for relatively small values of $n-m$. The idea is to provide a decomposition of the generating function $\widehat{F}_d$ into simpler rational fractions whose *inversion* is easy to obtain. To avoid handling complex roots of unity, we do not use residues *explicitly*, but build up the required decomposition in a recursive manner. Roughly speaking, we inductively compute real constants $Q_{\sigma,\beta}$ and a fixed positive integer M, all of them completely independent of y, such that the *counting* function $\widehat{f}_d$ is given by the finite sum

$$\widehat{f}_d(y,c) = \sum_{A_\sigma} \sum_{\beta \in \mathbb{Z}^m, \|\beta\| \leq M} Q_{\sigma,\beta} \times \begin{cases} e^{c'_\sigma x} & \text{if } x := A_\sigma^{-1}[y-\beta] \in \mathbb{N}^m \\ 0 & \text{otherwise,} \end{cases}$$

where the first finite sum is computed over all invertible $[m \times m]$-square submatrices A_σ of A. Crucial in our algorithm is an explicit decomposition in *closed form* (and thus an explicit formula for $\widehat{f}_d(y,c)$) for the case $n = m+1$, which we then repeatedly use for the general case $n > m+1$.

Preliminaries. With no loss of generality, we may and will suppose from now on that the matrix $A \in \mathbb{Z}^{m \times n}$ has maximal rank.

Definition 4.1. Let $p \in \mathbb{N}$ satisfy $m \leq p \leq n$, and let $\eta = \{\eta_1, \eta_2, \ldots, \eta_p\} \subset \mathbb{N}$ be an ordered set with cardinality $|\eta| = p$ and $1 \leq \eta_1 < \eta_2 < \cdots < \eta_p \leq n$. Then

(i) η is said to be a *basis* of order p if the $[m \times p]$ submatrix

$$A_\eta := \left[A_{\eta_1} | A_{\eta_2} | \cdots | A_{\eta_p}\right]$$

has maximal rank, that is, $\operatorname{rank}(A_\eta) = m$.

(ii) For $m \leq p \leq n$, let

$$\mathbf{J}_p := \{\eta \subset \{1, \ldots, n\} \mid \eta \text{ is a basis of order } p\} \tag{4.22}$$

be the set of bases of order p.

Notice that $\mathbf{J}_n = \{\{1,2,\ldots,n\}\}$ because A has maximal rank. Moreover,

Lemma 4.1. *Let η be any subset of $\{1,2,\ldots,n\}$ with cardinality $|\eta|$.*

(i) If $|\eta| = m$ then $\eta \in \mathbf{J}_m$ if and only if A_η is invertible.
(ii) If $|\eta| = q$ with $m < q \leq n$, then $\eta \in \mathbf{J}_q$ if and only if there exists a basis $\sigma \in \mathbf{J}_m$ such that $\sigma \subset \eta$.

Proof. (i) is immediate because A_η is a square matrix, and A_η is invertible if and only if A_η has maximal rank.

On the other hand, (ii) also follows from the fact that A_η has maximal rank if and only if A_η contains a square invertible submatrix. □

Lemma 4.1 automatically implies $\mathbf{J}_m \neq \emptyset$ because the matrix A must contain at least one square invertible submatrix (we are supposing that A has maximal rank). Besides, $\mathbf{J}_p \neq \emptyset$ for $m < p \leq n$, because $\mathbf{J}_m \neq \emptyset$.

Finally, given a basis $\eta \in \mathbf{J}_p$ for $m \leq p \leq n$, and three vectors $z \in \mathbb{C}^m$, $c \in \mathbb{R}^n$, and $w \in \mathbb{Z}^m$, we introduce the following notation .

$$\begin{aligned} c_\eta &:= (c_{\eta_1}, c_{\eta_2}, \ldots c_{\eta_p})', \\ \|w\| &:= \max\{|w_1|, |w_2|, \ldots, |w_m|\}. \end{aligned} \tag{4.23}$$

Definition 4.2. The vector $c \in \mathbb{R}^n$ is said to be *regular* if for every basis $\sigma \in \mathbf{J}_{m+1}$, there exists a nonzero vector $v(\sigma) \in \mathbb{Z}^{m+1}$ such that

$$A_\sigma v(\sigma) = 0 \quad \text{and} \quad c'_\sigma v(\sigma) \neq 0. \tag{4.24}$$

Notice that $c \neq 0$ whenever c is regular. Moreover, there are infinitely many vectors $v \in \mathbb{Z}^{m+1}$ such that $A_\sigma v = 0$, because $\operatorname{rank}(A_\sigma) = m < n$. Thus, the vector $c \in \mathbb{R}^n$ is regular if and only if

$$c_j - c'_\pi A_\pi^{-1} A_j \neq 0 \quad \forall \pi \in \mathbf{J}_m, \quad \forall j \notin \pi;$$

this is the regularity condition defined in Section 2.3 and required in Brion and Vergne's formula (4.14)–(4.15).

Let $c \in \mathbb{R}^n$ be such that $c < A'u$ for some $u \in \mathbb{R}^m$, and so the generating function $z \mapsto \widehat{F}_d(z,c)$ in (4.10) is well-defined; for convenience, redefine $\widehat{F}_d$ as

$$\widehat{F}_d(z,c) = \sum_{y \in \mathbb{Z}^m} \widehat{f}_d(y,c)\, z^y \tag{4.25}$$

$$= \prod_{k=1}^{n} \frac{1}{\prod_{j=1}^{n}(1 - \mathrm{e}^{c_j} z^{A_j})} \tag{4.26}$$

on the domain

$$|z|^A < \mathrm{e}^{-c}. \tag{4.27}$$

A decomposition of the generating function

We will compute the exact value of $\widehat{f}_d(y,c)$ by first determining an appropriate expansion of the generating function in the form

$$\widehat{F}_d(z,c) = \sum_{\sigma\in\mathbf{J}_m} \frac{Q_\sigma(z)}{\prod_{k\in\sigma}(1-\mathrm{e}^{c_k}z^{A_k})}, \tag{4.28}$$

where the coefficients $Q_\sigma : \mathbb{C}^m \to \mathbb{C}$ are *rational* functions with a finite Laurent series

$$z \mapsto Q_\sigma(z) = \sum_{\beta\in\mathbb{Z}^m,\,\|\beta\|\leq M.} Q_{\sigma,\beta} z^\beta. \tag{4.29}$$

In (4.29), the strictly positive integer M is fixed and each $Q_{\sigma,\beta}$ is a real constant.

Remark 4.1. The decomposition (4.28) is *not* unique (at all) and there are several ways to obtain such a decomposition. For instance, Brion and Vergne [27, §2.3, p. 815] provide an explicit decomposition of $\widehat{F}_d(z,c)$ into elementary rational fractions of the form

$$\widehat{F}_d(z,c) = \sum_{\sigma\in\mathbf{J}_m}\sum_{g\in G(\sigma)} \frac{1}{\prod_{j\in\sigma}\left(1-\gamma_j(g)(\mathrm{e}^{c_j}z^{A_j})^{1/q}\right)} \frac{1}{\prod_{k\notin\sigma}\delta_k(g)}, \tag{4.30}$$

where $G(\sigma)$ is a certain set of cardinality q, and the coefficients $\{\gamma_j(g),\delta_k(g)\}$ involve certain roots of unity. The fact that c is *regular* ensures that (4.30) is well-defined. Thus, in principle, we could obtain (4.28) from (4.30), but this would require a highly nontrivial analysis and manipulation of the coefficients $\{\gamma_j(g),\delta_k(g)\}$. In the sequel, we provide an alternative algebraic approach that avoids manipulating these complex coefficients.

If $\widehat{F}_d$ satisfies (4.28) then we get the following result:

Theorem 4.2. *Let $A\in\mathbb{Z}^{m\times n}$ be of maximal rank, $\widehat{F}_d$ be as in (4.25) with $-c < A'u$ for some $u\in\mathbf{K}$, and assume that the generating function $\widehat{F}_d$ satisfies (4.28)–(4.29). Then*

$$\widehat{f}_d(y,c) = \sum_{\sigma\in\mathbf{J}_m}\ \sum_{\beta\in\mathbb{Z}^m,\,\|\beta\|\leq M} Q_{\sigma,\beta}\, E_\sigma(y-\beta), \tag{4.31}$$

with

$$E_\sigma(y-\beta) = \begin{cases} \mathrm{e}^{c'_\sigma x} & \text{if } x := A_\sigma^{-1}[y-\beta]\in\mathbb{N}^m, \\ 0 & \text{otherwise,} \end{cases} \tag{4.32}$$

where $c_\sigma\in\mathbb{R}^m$ is defined in (4.23).

Proof. Recall that $z^{A_k} = z_1^{A_{1k}}\cdots z_m^{A_{mk}}$. On the other hand, $|\mathrm{e}^{c_k}z^{A_k}| < 1$ for all $1\leq k\leq n$. Therefore, for each $\sigma\in\mathbf{J}_m$, one has the expansion

$$\prod_{k\in\sigma}\frac{1}{1-\mathrm{e}^{c_k}z^{A_k}} = \prod_{k\in\sigma}\left[\sum_{x_k\in\mathbb{N}} \mathrm{e}^{c_k x_k} z^{A_k x_k}\right] = \sum_{x\in\mathbb{N}^m} \mathrm{e}^{c'_\sigma x} z^{A_\sigma x}.$$

Next, suppose that a decomposition (4.28)–(4.29) exists. Then one gets the relationship

$$\begin{aligned}\widehat{F}_d(z,c) &= \sum_{\sigma\in\mathbf{J}_m} \sum_{x\in\mathbb{N}^m} Q_\sigma(z)\, \mathrm{e}^{c'_\sigma x} z^{A_\sigma x} \\ &= \sum_{\sigma\in\mathbf{J}_m} \sum_{\beta\in\mathbb{Z}^m, \|\beta\|\le M} \sum_{x\in\mathbb{N}^m} Q_{\sigma,\beta}\, \mathrm{e}^{c'_\sigma x} z^{\beta+A_\sigma x}. \end{aligned} \tag{4.33}$$

Notice that (4.25) and (4.33) are the same. Hence, if we want to obtain the exact value of $\widehat{f}_d(y,c)$ from (4.33), we only have to sum up all the terms whose exponent $\beta + A_\sigma x$ is equal to y. That is, recalling that A_σ is invertible for every $\sigma \in \mathbf{J}_m$ (see Lemma 4.1),

$$\widehat{f}_d(y,c) = \sum_{\sigma\in\mathbf{J}_m} \sum_{\beta\in\mathbb{Z}^m, \|\beta\|\le M} Q_{\sigma,\beta} \times \begin{cases} \mathrm{e}^{c'_\sigma x} & \text{if } x := A_\sigma^{-1}[y-\beta] \in \mathbb{N}^m, \\ 0 & \text{otherwise,} \end{cases}$$

which is exactly (4.31). □

In view of Theorem 4.2, $\widehat{f}_d$ is easily obtained once the rational functions $Q_\sigma(z)$ in the decomposition (4.28) are available. As already pointed out, the decomposition (4.28)–(4.29) is not unique and in what followgraph we provide a simple decomposition (4.28) for which the coefficients Q_σ are easily calculated in the case $n = m+1$, and a recursive algorithm to provide the Q_σ in the general case $n > m+1$.

The case $\mathbf{n = m+1}$. Here we completely solve the case $n = m+1$, that is, we provide an explicit expression of $\widehat{f}_d$.

Let $\mathrm{sgn} : \mathbb{R}\to\mathbb{Z}$ be the *sign* function, i.e.,

$$t \mapsto \mathrm{sgn}(t) := \begin{cases} 1 & \text{if } \; t > 0, \\ -1 & \text{if } \; t < 0, \\ 0 & \text{otherwise.} \end{cases}$$

In addition, we adopt the convention $\sum_{r=0}^{-1}(\ldots)$ is identically equal to zero.

Given a fixed integer $q > 0$ and for every $k = 1,\ldots, n$, we are going to construct auxiliary functions $P_k : \mathbb{Z}^n \times \mathbb{C}^n \to \mathbb{C}$ such that each $w \mapsto P_k(\theta, w)$ is a rational function of the variable $w \in \mathbb{C}^n$. Given a vector $v \in \mathbb{Z}^n$, we define

$$\begin{aligned} P_1(v,w) &:= \sum_{r=0}^{|v_1|-1} w_1^{\mathrm{sgn}(v_1)\,r}, \\ P_2(v,w) &:= \left[w_1^{v_1}\right] \sum_{r=0}^{|v_2|-1} w_2^{\mathrm{sgn}(v_2)\,r}, \\ P_3(v,w) &:= \left[w_1^{v_1} w_2^{v_2}\right] \sum_{r=0}^{|v_3|-1} w_3^{\mathrm{sgn}(v_3)\,r}, \\ &\;\;\vdots \qquad\qquad \vdots \\ P_n(v,w) &:= \left[\prod_{j=1}^{n-1} w_j^{v_j}\right] \sum_{r=0}^{|v_n|-1} w_q^{\mathrm{sgn}(v_n)\,r}. \end{aligned} \tag{4.34}$$

Obviously, $P_k(v,w)=0$ whenever $v_k=0$. Moreover,

Lemma 4.2. *Let $v\in\mathbb{Z}^n$ and $w\in\mathbb{C}^n$. The functions P_k defined in (4.34) satisfy*

$$\sum_{k=1}^{n}\left(1-w_k^{\mathrm{sgn}(v_k)}\right)P_k(v,w)=1-w^v. \tag{4.35}$$

Proof. First, notice that

$$\left(1-w_1^{\mathrm{sgn}(v_1)}\right)P_1(v,w)=\left(1-w_1^{\mathrm{sgn}(v_1)}\right)\sum_{r=0}^{|v_1|-1}w_1^{\mathrm{sgn}(v_1)r}=1-w_1^{v_1}.$$

The above equality is obvious when $v_1=0$. Similar formulas hold for $2\le k\le n$; namely

$$\left(1-w_k^{\mathrm{sgn}(v_k)}\right)P_k(v,w)=(1-w_k^{v_k})\prod_{j=1}^{k-1}w_j^{v_j}=\prod_{j=1}^{k-1}w_j^{v_j}-\prod_{j=1}^{k}w_j^{v_j}.$$

Therefore, summing up in (4.35) yields

$$\sum_{k=1}^{n}\left(1-w_k^{\mathrm{sgn}(v_k)}\right)P_k(v,w)=1-\prod_{j=1}^{n}w_j^{v_j}.$$

□

Solving for the case $\mathbf{n}=\mathbf{m}+\mathbf{1}$. We now use P_k to evaluate $\widehat{f}_d(y,c)$ when $A\in\mathbb{Z}^{m\times(m+1)}$ is a maximal rank matrix.

Theorem 4.3. *Let $n=m+1$ be fixed, $A\in\mathbb{Z}^{m\times n}$ be a maximal rank matrix, and let c be regular and as in Theorem 4.2. Let $v\in\mathbb{Z}^n$ be a nonzero vector such that $Av=0$ and $c'v\neq 0$ (see Definition 4.2). Define the vector*

$$w:=(e^{c_1}z^{A_1},e^{c_2}z^{A_2},\ldots,e^{c_n}z^{A_n}). \tag{4.36}$$

Then

(i) The generating function $\widehat{F}_d(z,c)$ has the expansion

$$\widehat{F}_d(z,c)=\sum_{k=1}^{n}\frac{Q_k(z)}{\prod_{j\neq j_k}(1-\mathrm{e}^{c_j}z^{A_j})}=\sum_{\sigma\in\mathbf{J}_m}\frac{Q_\sigma(z)}{\prod_{j\in\sigma}(1-\mathrm{e}^{c_j}z^{A_j})}, \tag{4.37}$$

where the rational functions $z\mapsto Q_k(z)$ are defined by

$$Q_k(z):=\begin{cases}P_k(v,w)/(1-\mathrm{e}^{c'v}) & \text{if } v_k>0,\\ -w_k^{-1}P_k(v,w)/(1-\mathrm{e}^{c'v}) & \text{if } v_k<0,\\ 0 & \text{otherwise}\end{cases} \tag{4.38}$$

for $1 \leq k \leq n$. Each function P_k in (4.38) is defined as in (4.34). (Notice that $Q_k = 0$ if $v_k = 0$.)

(ii) Given $y \in \mathbb{Z}^m$, the function $\widehat{f_d}(y,c)$ is directly obtained by application of Theorem 4.2.

Remark 4.2. (a) In the case where $n = m+1$ and $\Omega(y)$ is compact, a naive way to evaluate $\widehat{f_d}(y,c)$ is as follows. Suppose that $B := [A_1|\cdots|A_m]$ is invertible. One may then calculate $\rho := \max\{x_{m+1} \,|\, Ax = y, x \geq 0\}$. Thus, evaluating $\widehat{f_d}(y,c)$ reduces to summing up $\sum_x e^{c'x}$ over all vectors $x = (\hat{x}, x_{m+1}) \in \mathbb{N}^{m+1}$ such that $x_{m+1} \in [0,\rho] \cap \mathbb{N}$ and $\hat{x} := B^{-1}[y - A_{m+1}x_{m+1}]$. This may work very well for reasonable values of ρ; but clearly, ρ depends on the magnitude of y. On the other hand, the computational complexity of evaluation of $\widehat{f_d}(y,c)$ via (4.31) does *not* depend on y. Indeed, the bound M in (4.31) of Theorem 4.2 does not depend at all on y. Moreover, the method also works if $\Omega(y)$ is not compact.

To illustrate the difference, consider the following trivial example, where $n = 2, m = 1, A = [1,1]$, and $c = [0,a]$ with $a \neq 0$. The generating function $\widehat{F_d}(z,c)$ is the rational function

$$\widehat{F_d}(z,c) = \frac{1}{(1-z)(1-e^a z)}.$$

Setting $v = (-1,1)$ and $w = (z, e^a z)$, one obtains

$$\begin{aligned} 1 &= (1-z)\,Q_1(z) + (1-e^a z)\,Q_2(z) \\ &= (1-z)\frac{-z^{-1}P_1(v,w)}{1-e^a} + (1-e^a z)\frac{P_2(v,w)}{1-e^a} \\ &= (1-z)\frac{-z^{-1}}{1-e^a} + (1-e^a z)\frac{z^{-1}}{1-e^a}, \end{aligned}$$

an illustration of the Hilbert's Nullstellensatz applied to the two polynomials $z \mapsto 1-z$ and $z \mapsto 1-e^a z$, which have no common zero in $\mathbb{C}$.

And so, the generating function $\widehat{F_d}(z,c)$ gets expanded to

$$\widehat{F_d}(z,c) = \frac{-z^{-1}}{(1-e^a)(1-e^a z)} + \frac{z^{-1}}{(1-e^a)(1-z)}. \tag{4.39}$$

Finally, using Theorem 4.2, we obtain $\widehat{f_d}(y,c)$ in closed form by

$$\widehat{f_d}(y,c) = \frac{-e^{a(y+1)}}{1-e^a} + \frac{e^{0(y+1)}}{1-e^a} = \frac{1-e^{(y+1)a}}{1-e^a}. \tag{4.40}$$

Looking back at (4.29) we may see that $M = 1$ (which obviously does not depend on y) and so, evaluating $\widehat{f_d}(y,c)$ via (4.31) (i.e., as in (4.40)) is done in two elementary steps, no matter the magnitude of y. On the other hand, the naive procedure would require y elementary steps.

(b) We have already mentioned that the expansion of the generating function $\widehat{F_d}(z,c)$ is not unique. In the above trivial example, we also have

$$\widehat{F}_d(z,c) = \frac{\mathrm{e}^a}{(\mathrm{e}^a-1)(1-\mathrm{e}^a z)} - \frac{1}{(\mathrm{e}^a-1)(1-z)},$$

which is not the same expansion as (4.39). However, applying Theorem 4.2 again yields the same formula (4.40) for $\widehat{f}_d(y,c)$.

The general case *n* > *m*+*1*. We now consider the case $n > m+1$ and obtain a decomposition (4.28) that permits us to compute $\widehat{f}_d(y,c)$ by invoking Theorem 4.2. The idea is to use recursively the results for the case $n = m+1$ so as to exhibit a decomposition (4.28) in the general case $n > m+1$ by induction.

Proposition 4.2. *Let $A \in \mathbb{Z}^{m\times n}$ be a maximal rank matrix, c be regular and as in Theorem 4.2. Suppose that the generating function $\widehat{F}_d$ has the expansion*

$$\widehat{F}_d(z,c) = \sum_{\pi\in\mathbf{J}_p} \frac{Q_\pi(z)}{\prod_{k\in\pi}(1-\mathrm{e}^{c_k}z^{A_k})}, \tag{4.41}$$

for some integer p with $m < p \leq n$, and for some rational functions $z \mapsto Q_\pi(z)$, explicitly known, and with the finite Laurent's series expansion (4.29). Then, $\widehat{F}_d$ also has the expansion

$$\widehat{F}_d(z,c) = \sum_{\breve{\pi}\in\mathbf{J}_{p-1}} \frac{Q^*_{\breve{\pi}}(z)}{\prod_{k\in\breve{\pi}}(1-\mathrm{e}^{c_k}z^{A_k})}, \tag{4.42}$$

where the rational functions $z \mapsto Q^*_{\breve{\pi}}(z)$ are constructed explicitly and have a finite Laurent's series expansion (4.29).

Proof. Let $\pi \in \mathbf{J}_p$ be a given basis with $m < p \leq n$ and such that $Q_\pi(z) \not\equiv 0$ in (4.41). We are going to build up simple rational functions $z \mapsto R^\pi_\eta(z)$, where $\eta \in \mathbf{J}_{p-1}$, such that the expansion

$$\frac{1}{\prod_{j\in\pi}(1-\mathrm{e}^{c_j}z^{A_j})} = \sum_{\eta\in\mathbf{J}_{p-1}} \frac{R^\pi_\eta(z)}{\prod_{j\in\eta}(1-\mathrm{e}^{c_j}z^{A_j})} \tag{4.43}$$

holds. Invoking Lemma 4.1, there exists a basis $\breve{\sigma} \in \mathbf{J}_m$ such that $\breve{\sigma} \subset \pi$. Pick up any index $g \in \pi\setminus\breve{\sigma}$, so the basis $\sigma := \breve{\sigma}\cup\{g\}$ is indeed an element of $\mathbf{J}_{m+1}$, because of Lemma 4.1 again. Next, since c is regular, pick up a vector $v \in \mathbb{Z}^{m+1}$ such that $A_\sigma v = 0$ and $c'_\sigma v \neq 0$, like in (4.24). The statements below follow from the same arguments as in the proof of Theorem 4.3(i), so we only sketch the proof. Define the vector

$$w_\sigma := (\mathrm{e}^{c_1}z^{A_1}, \mathrm{e}^{c_2}z^{A_2}, \ldots, \mathrm{e}^{c_n}z^{A_n}) \in \mathbb{C}^n, \tag{4.44}$$

and so $w_\sigma \in \mathbb{C}^{m+1}$. One easily deduces that $w^v_\sigma = \mathrm{e}^{c'_\sigma v} \neq 1$. Moreover, define the rational functions

$$R^\pi_k(z) := \begin{cases} P_k(v,w_\sigma)/(1-\mathrm{e}^{c'_\sigma v}) & \text{if } v_k > 0, \\ -[w_\sigma]^{-1}_k P_k(v,w_\sigma)/(1-\mathrm{e}^{c'_\sigma v}) & \text{if } v_k < 0, \\ 0 & \text{otherwise,} \end{cases} \tag{4.45}$$

where the functions P_k are defined as in (4.34), for all $1 \le k \le m+1$. Therefore,

$$1 = \sum_{k=1}^{m+1} (1 - [w_\sigma]_k) \, R_k^\pi(z) = \sum_{k=1}^{m+1} \left(1 - e^{c_{\sigma_k}} z^{A_{\sigma_k}}\right) R_k^\pi(z).$$

The latter implies

$$\frac{1}{\prod_{j\in\pi}(1 - e^{c_j} z^{A_j})} = \sum_{k=1}^{q} R_k^\pi(z) \left[\prod_{j\in\pi,\, j\neq\sigma_k} \frac{1}{1 - e^{c_j} z^{A_j}} \right]. \tag{4.46}$$

Next, we use the same arguments as in the proof of Theorem 4.3(ii).

With no loss of generality, suppose that the ordered sets $\breve{\sigma} \subset \sigma \subset \pi$ are given by

$$\pi := \{1, 2, \ldots, p\}, \quad \sigma := \breve{\sigma} \cup \{p\}, \quad \text{and} \quad p \notin \breve{\sigma}. \tag{4.47}$$

Notice that $\sigma_k \in \breve{\sigma}$ for $1 \le k \le m$, and $\sigma_{m+1} = p$. In addition, consider the ordered sets

$$\eta(k) = \{ j \in \pi \mid j \neq \sigma_{j_k} \} \quad \text{for } k = 1, \ldots, m+1. \tag{4.48}$$

We next show that each submatrix $A_{\eta(k)}$ has maximal rank for every $k = 1, \ldots, m+1$ with $v_k \neq 0$. Notice that $|\eta(k)| = p - 1$ because $|\pi| = p$; hence, the set $\eta(k)$ is indeed an element of $\mathbf{J}_{p-1}$ precisely when $A_{\eta(k)}$ has maximal rank. Now $\breve{\sigma} \subset \eta(m+1)$ because $\sigma_{m+1} = p$ is contained in $\pi \setminus \breve{\sigma}$. Therefore, since $\breve{\sigma} \in \mathbf{J}_m$, Lemma 4.1 implies that $\eta(m+1)$ in (4.48) is an element of $\mathbf{J}_{p-1}$, the square submatrix $A_{\breve{\sigma}}$ is invertible, and $A_{\eta(m+1)}$ has maximal rank. On the other hand, the vector $v \in \mathbb{Z}^{m+1}$ satisfies

$$0 = A_\sigma v = A_{\breve{\sigma}} (v_1, v_2, \ldots, v_m)' + A_p v_{m+1},$$

with $v \neq 0$, and so we may conclude that $v_{m+1} \neq 0$. We can now express the pth column of A as the finite sum $A_p = \sum_{j=1}^{m} \frac{-v_j}{v_{m+1}} A_{\sigma_j}$. Hence, for every $1 \le k \le m$ with $v_k \neq 0$, the matrix

$$A_{\eta(k)} = \left[A_1 | \cdots | A_{\sigma_k - 1} | A_{\sigma_k + 1} | \cdots | A_p \right]$$

has maximal rank, because the column A_{σ_k} of $A_{\sigma(m+1)}$ has been substituted with the linear combination $A_p = \sum_{j=1}^{m} \frac{-v_j}{v_{m+1}} A_{\sigma_j}$ whose coefficient $-v_k/v_{m+1}$ is different from zero. Thus, the set $\eta(k)$ defined in (4.48) is an element of $\mathbf{J}_{p-1}$ for every $1 \le k \le m$ with $v_k \neq 0$. Then expansion (4.46) can be rewritten as

$$\begin{aligned} \frac{1}{\prod_{j\in\pi}(1 - e^{c_j} z^{A_j})} &= \sum_{v_k \neq 0} \frac{R_k^\pi(z)}{\prod_{j\in\eta(k)}(1 - e^{c_j} z^{A_j})} \\ &= \sum_{\eta \in \mathbf{J}_{p-1}} \frac{R_\eta^\pi(z)}{\prod_{j\in\eta}(1 - e^{c_j} z^{A_j})}, \end{aligned}$$

which is the desired identity (4.43) with

$$R^{\pi}_{\eta}(z) \equiv \begin{cases} R^{\pi}_{k}(z) & \text{if } \eta = \eta(k) \text{ for some index } k \text{ with } v_k \neq 0, \\ 0 & \text{otherwise.} \end{cases}$$

On the other hand, it is easy to see that all rational functions R^{π}_k and R^{π}_{η} have finite Laurent series (4.29), because each R^{π}_k is defined in terms of P_k in (4.45), and each rational function P_k defined in (4.34) has a finite Laurent series. Finally, combining (4.41) and (4.43) yields

$$\widehat{F}_d(z,c) = \sum_{\eta \in \mathbf{J}_{p-1}} \sum_{\pi \in \mathbf{J}_p} \frac{R^{\pi}_{\eta}(z)\, Q_{\pi}(z)}{\prod_{k\in\eta}(1-\mathrm{e}^{c_k} z^{A_k})}, \tag{4.49}$$

so that the decomposition (4.42) holds by setting Q^*_{η} identically equal to the finite sum $\sum_{\pi\in\mathbf{J}_p} R^{\pi}_{\eta} Q_{\pi}$ for every $\eta \in \mathbf{J}_{p-1}$. □

Notice that the sum in (4.41) runs over the bases of order p, whereas the sum in (4.42) runs over the bases of order $p-1$. Hence, repeated applications of Proposition 4.2 yield a decomposition of the generating function $\widehat{F}_d$ into a sum over the bases of order m, which is the decomposition described in (4.28)–(4.29). Namely,

Corollary 4.1. *Let $A \in \mathbb{Z}^{m\times n}$ be a maximal rank matrix and c be regular and as in Theorem 4.2. Let $\widehat{f}_d$ be as in (4.7) and $\widehat{F}_d$ be its generating function (4.25)–(4.26). Then*

(i) $\widehat{F}_d(z,c)$ has the expansion

$$\widehat{F}_d(z,c) = \sum_{\sigma\in\mathbf{J}_m} \frac{Q_{\sigma}(z)}{\prod_{k\in\sigma}(1-\mathrm{e}^{c_k} z^{A_k})}, \tag{4.50}$$

for some rational functions $z \mapsto Q_{\sigma}(z)$, which can be built up explicitly, and with finite Laurent series (4.29).

(ii) For every $y \in \mathbb{Z}^m$, the function $\widehat{f}_d(y,c)$ is obtained by direct application of Theorem 4.2.

Proof. Point (i) is proved by induction. Notice that (4.26) can be rewritten as

$$\widehat{F}_d(z,c) = \sum_{\pi\in\mathbf{J}_n} \frac{1}{\prod_{k\in\pi}(1-\mathrm{e}^{c_k} z^{A_k})},$$

because $\mathbf{J}_n = \{\{1,2,\ldots,n\}\}$ and A has maximal rank (see (4.22)). Thus, from Proposition 4.2, (4.42) holds for $p = n-1$ as well. And more generally, repeated applications of Proposition 4.2 show that (4.42) holds for all $m \leq p < n$. However, (4.50) is precisely (4.42) with $p = m$.

On the other hand, (ii) follows because in view of our hypothesis on c, $z \mapsto \widehat{F}_d(z,c)$ is the generating function of $y \mapsto \widehat{f}_d(y,c)$ and has the decomposition (4.50) required to apply Theorem 4.2. □

An algorithm for the general case $n > m+1$

Let $S[x]$ be the set of rational functions with *finite* Laurent series. That is, $Q \in S[x]$ if (4.29) holds. In view of Corollary 4.1, the recursive algorithm for the general case reads as follows:

Algorithm 1 to compute $\widehat{f}_d(y,c)$:
Input: $m, n > m+1$; $\gamma := e^c \in \mathbb{R}^n, y \in \mathbb{Z}^m$; $A \in \mathbb{Z}^{m\times n}$ full rank.
Output: $\widehat{f}_d(y,c)$ as in (4.31).
Step 0: Initialization. $p := n$. $\mathbf{J}_p := \{\{1,2,\ldots,n\}\}$; $\pi := \{1,2,\ldots,n\} \in \mathbf{J}_p$; $z \mapsto Q_\pi(z) \equiv 1 \in S[x]$.

$$\widehat{F}_d(z,c) = \sum_{\pi \in \mathbf{J}_p} \frac{Q_\pi(z)}{\prod_{j\in\pi}(1-e^{c_j}z^{A_j})}.$$

Step 1: While $p \geq m+1$ proceed as follows:

- For all $\pi \in \mathbf{J}_p$ with $Q_\pi \not\equiv 0$,
  (a) pick a basis $\breve{\sigma} \in \mathbf{J}_m$ and pick $g \in \pi \setminus \breve{\sigma}$
  (b) let $\sigma := \breve{\sigma} \cup \{g\} \in \mathbf{J}_{m+1}$
  (c) let $A_\sigma := [A_j]_{j\in\sigma} \in \mathbb{Z}^{m\times(m+1)}, c_\sigma := [c_j]_{j\in\sigma} \in \mathbb{R}^{m+1}$
  (d) `solve`(A_σ, c_σ, y) yielding $\{R_k^\pi\}_{k\in\sigma} \subset S[x]$ so that

$$\frac{1}{\prod_{j\in\pi}(1-e^{c_j}z^{A_j})} = \sum_{k\in\sigma} \frac{R_k^\pi(z)}{\prod_{j\in\pi\setminus\{k\}}(1-e^{c_j}z^{A_j})}$$

  (e) let $\pi_k := \pi \setminus \{k\} \in \mathbf{J}_{p-1}$ and $Q_{\pi_k} := Q_\pi R_k^\pi \in S[x]$ for all $k \in \sigma$. Then

$$\frac{Q_\pi(z)}{\prod_{j\in\pi}(1-e^{c_j}z^{A_j})} = \sum_{k\in\sigma} \frac{Q_{\pi_k}(z)}{\prod_{j\in\pi_k}(1-e^{c_j}z^{A_j})}.$$

- We have now

$$\widehat{F}_d(z,c) = \sum_{\eta\in\mathbf{J}_{p-1}} \frac{Q_\eta(z)}{\prod_{j\in\eta}(1-e^{c_j}z^{A_j})}, \quad \text{with } Q_\eta \in S[x], \forall \eta \in \mathbf{J}_{p-1}.$$

- Set $p := p-1$.

Step 2: $p = m$ and

$$\widehat{F}_d(z,c) = \sum_{\pi\in\mathbf{J}_m} \frac{Q_\pi(z)}{\prod_{j\in\pi}(1-e^{c_j}z^{A_j})}, \quad \text{with } Q_\pi \in S[x], \forall \pi \in \mathbf{J}_m,$$

that is, $\widehat{F}_d(z,c)$ is in the form (4.28) required to apply Theorem 4.2. Thus, get $\widehat{f}_d(y,c)$ from (4.31) in Theorem 4.2.

In Step 1(d) of the above algorithm, `Solve`(A_σ, c_σ, y) is just application of Theorem 4.3 with $A = A_\sigma$ and $c = c_\sigma$.

Computational complexity

First, observe that the above algorithm computes the coefficients of the polynomials $Q_\sigma(z)$ in the decomposition (4.28) of $\widehat{F}_d(z,c)$. This computation involves only simple linear algebra operations, provided that the matrix $A \in \mathbb{Z}^{m\times n}$ and the vector $c \in \mathbb{R}^n$ are given. Recall that an important step in these procedures is to compute a vector $v \in \mathbb{Z}^{m+1}$ such that $A_\sigma v = 0$ and $c'_\sigma v \neq 0$. Thus, for practical implementation, one should directly consider working with a rational vector $c \in \mathbb{Q}^n$. Next, one may easily see that in the algorithm, the entries e^{c_k} can be treated symbolically. Indeed, we need the numerical value of each e^{c_k} only at the very final step, i.e., for evaluating $\widehat{f}_d(y,c)$ via (4.31) in Theorem 4.2; see the illustrative example, Example 4.3. Therefore, only at the very final stage, one needs a good rational approximation of e^{c_j} in $\mathbb{Q}^n$ to evaluate $\widehat{f}_d(y,c)$.

Having said this, the computational complexity is essentially determined by the number of coefficients $\{Q_{\sigma,\beta}\}$ in equation (4.31); or equivalently, by the number of nonzero coefficients of the polynomials $\{Q_\sigma(z)\}$ in the decomposition (4.28)–(4.29). Define

$$\Lambda := \max_{\sigma\in\mathbf{J}_{m+1}} \left\{ \min\{ \|v\| \,|\, A_\sigma v = 0,\ c'_\sigma v \neq 0,\ v \in \mathbb{Z}^{m+1}\} \right\}. \tag{4.51}$$

In the case $n = m+1$, each polynomial $Q_\sigma(z)$ has at most Λ terms. This follows directly from (4.34) and (4.38).

For $n = m+2$, we have at most $(m+1)^2$ polynomials $Q_\sigma(z)$ in (4.28); and again, each one of them has at most Λ nonzero coefficients. Therefore, in the general case $n > m$, we end up with at most $(m+1)^{n-m}\Lambda$ terms in (4.31). Thus, the computational complexity is equal to $\mathrm{O}[(m+1)^{n-m}\Lambda]$, which makes it particularly attractive for relatively small values of $n-m$.

As a nice feature of the algorithm, notice that the computational complexity does *not* depend on the right-hand side $y \in \mathbb{Z}^m$. Moreover, notice that the constant Λ does not change (at all) if we multiply the vector $c \in \mathbb{Q}^n$ for any real $r \neq 0$, because $c'_\sigma v \neq 0$ if and only if $rc'_\sigma v \neq 0$. Hence, we can also conclude that the computational complexity does *not* depend on the magnitude of $\|c\|$, it only depends on the ratio between the entries of c. However, as shown in the following simple example, Λ is exponential in the input size of A. Indeed, if

$$A = \begin{bmatrix} 1 & a & a^2 \\ 1 & a+1 & (a+1)^2 \end{bmatrix},$$

then necessarily every solution $v \in \mathbb{Z}^3$ of $Av = 0$ is an integer multiple of the vector $(a^2+a, -2a-1, 1)$, and so $\Lambda = O(a^2)$. Finally, the constant $M > 0$ in (4.29) and (4.31) depends polynomially on Λ.

Example 4.3. Consider the following example with $n = 6, m = 3$, and data

$$A := \begin{bmatrix} 1\,1\,1\,1\,0\,0 \\ 2\,1\,0\,0\,1\,0 \\ 0\,2\,1\,0\,0\,1 \end{bmatrix}, \quad c := (c_1,\ldots,c_6),$$

so that $\widehat{F}(z,c)$ is the rational fraction

$$\frac{1}{(1-e^{c_1}z_1z_2^2)(1-e^{c_2}z_1z_2z_3^2)(1-e^{c_3}z_1z_3)(1-e^{c_4}z_1)(1-e^{c_5}z_2)(1-e^{c_6}z_3)}.$$

First Step: Setting $\pi = \{1,2,\ldots,6\} \in \mathbf{J}_6$, choose $\breve{\sigma} := \{4,5,6\}$ and $\sigma := \{3,4,5,6\}$. Let $v := (-1,1,0,1) \in \mathbb{Z}^4$ and solve $A_\sigma v = 0$. We obviously have $q = 3$, $\theta = (-1,1,1)$, $w = (\mathrm{e}^{c_3}z_1z_3, \mathrm{e}^{c_4}z_1, \mathrm{e}^{c_6}z_3)$, and so we get

$$R_1^\pi(z) = \frac{-(\mathrm{e}^{c_3}z_1z_3)^{-1}}{1-\mathrm{e}^{c_4+c_6-c_3}}; R_2^\pi(z) = \frac{(\mathrm{e}^{c_3}z_1z_3)^{-1}}{1-\mathrm{e}^{c_4+c_6-c_3}}; R_3^\pi(z) = \frac{\mathrm{e}^{(c_4-c_3)}z_3^{-1}}{1-\mathrm{e}^{c_4+c_6-c_3}}.$$

Hence

$$1 = (1-\mathrm{e}^{c_3}z_1z_3)R_1^\pi(z) + (1-\mathrm{e}^{c_4}z_1)R_2^\pi(z) + (1-\mathrm{e}^{c_6}z_3)R_3^\pi(z).$$

Notice that the term $(1-\mathrm{e}^{c_3}z_1z_3)R_1^\pi(z)$ will *annihilate* the element 3 in the base π. Moreover, the terms $(\cdots)R_2^\pi(z)$ and $(\cdots)R_3^\pi(z)$ will also *annihilate* the respective entries 4 and 6 in the base π, so

$$\begin{aligned}
\widehat{F}_d(z,c) &= \frac{R_1^\pi(z)}{(1-\mathrm{e}^{c_1}z_1z_2^2)(1-\mathrm{e}^{c_2}z_1z_2z_3^2)(1-\mathrm{e}^{c_4}z_1)(1-\mathrm{e}^{c_5}z_2)(1-\mathrm{e}^{c_6}z_3)} \\
&\quad + \frac{R_2^\pi(z)}{(1-\mathrm{e}^{c_1}z_1z_2^2)(1-\mathrm{e}^{c_2}z_1z_2z_3^2)(1-\mathrm{e}^{c_3}z_1z_3)(1-\mathrm{e}^{c_5}z_2)(1-\mathrm{e}^{c_6}z_3)} \\
&\quad + \frac{R_3^\pi(z)}{(1-\mathrm{e}^{c_1}z_1z_2^2)(1-\mathrm{e}^{c_2}z_1z_2z_3^2)(1-\mathrm{e}^{c_3}z_1z_3)(1-\mathrm{e}^{c_4}z_1)(1-\mathrm{e}^{c_5}z_2)} \\
&= \sum_{j=1}^{3} \frac{Q_{\eta_5(j)}(z)}{\prod_{k\in\eta_5(j)}(1-\mathrm{e}^{c_k}z^{A_k})},
\end{aligned}$$

where $\eta_5(1) = \{1,2,4,5,6\}$, $\eta_5(2) = \{1,2,3,5,6\}$, $\eta_5(3) = \{1,2,3,4,5\}$, and $Q_{\eta_5(j)}(z) = R_j^\pi(z)$ for $j = 1,2,3$.

Second Step: Analyzing $\eta_5(1) = \{1,2,4,5,6\} \in \mathbf{J}_5$, choose $\breve{\sigma} = \{4,5,6\}$ and $\sigma := \{1,4,5,6\}$. Let $v := (-1,1,2,0) \in \mathbb{Z}^4$ and solve $A_\sigma v = 0$. Then $q = 3$, $\theta = (-1,1,2)$, $w = (\mathrm{e}^{c_1}z_1z_2^2, \mathrm{e}^{c_4}z_1, \mathrm{e}^{c_5}z_2)$, and so we get

$$R_1^{\eta_5(1)}(z) = \frac{-(\mathrm{e}^{c_1}z_1z_2^2)^{-1}}{1-\mathrm{e}^{-c_1+c_4+2c_5}}, \quad R_2^{\eta_5(1)}(z) = \frac{(\mathrm{e}^{c_1}z_1z_2^2)^{-1}}{1-\mathrm{e}^{-c_1+c_4+2c_5}},$$

and

$$R_3^{\eta_5(1)}(z) = \frac{(\mathrm{e}^{c_4-c_1}z_2^{-2})(1+\mathrm{e}^{c_5}z_2)}{1-\mathrm{e}^{-c_1+c_4+2c_5}}.$$

Notice that the terms associated with $R_1^{\eta_5(1)}(z)$, $R_2^{\eta_5(1)}(z)$, and $R_3^{\eta_5(1)}(z)$ *annihilate* the respective entries 1, 4, and 5 in the base $\eta_5(1)$.

Analyzing $\eta_5(2) = \{1,2,3,5,6\} \in \mathbf{J}_5$, choose $\breve{\sigma} = \{3,5,6\}$ and $\sigma := \{2,3,5,6\}$. Let $v := (-1,1,1,1) \in \mathbb{Z}^4$ and solve $A_\sigma v = 0$. We have that $q = 4$, $\theta = v$, and $W = (\mathrm{e}^{c_2}z_1z_2z_3^2, \mathrm{e}^{c_3}z_1z_3, \mathrm{e}^{c_5}z_2, \mathrm{e}^{c_6}z_3)$, so we get

$$R_1^{\eta_5(2)}(z) = \frac{-(\mathrm{e}^{c_2}z_1z_2z_3^2)^{-1}}{1-\mathrm{e}^{-c_2+c_3+c_5+c_6}}, \quad R_2^{\eta_5(2)}(z) = \frac{(\mathrm{e}^{c_2}z_1z_2z_3^2)^{-1}}{1-\mathrm{e}^{-c_2+c_3+c_5+c_6}},$$

and

$$R_3^{\eta_5(2)}(z) = \frac{\mathrm{e}^{c_3-c_2}z_2^{-1}z_3^{-1}}{1-\mathrm{e}^{-c_2+c_3+c_5+c_6}}, \quad R_4^{\eta_5(2)}(z) = \frac{\mathrm{e}^{c_5+c_3-c_2}z_3^{-1}}{1-\mathrm{e}^{-c_2+c_3+c_5+c_6}}.$$

Notice that the terms associated to $R_1^{\eta_5(2)}(z)$, $R_2^{\eta_5(2)}(z)$, $R_3^{\eta_5(2)}(z)$, and $R_4^{\eta_5(2)}(z)$ *annihilate* the respective entries 2, 3, 5, and 6 in the base $\eta_5(2)$.

Analyzing $\eta_5(3) = \{1,2,3,4,5\} \in \mathbf{J}_5$, choose $\breve{\sigma} = \{3,4,5\}$ and $\sigma := \{2,3,4,5\}$. Let $v := (-1,2,-1,1) \in \mathbb{Z}^4$ and solve $A_\sigma v = 0$. We have that $q = 4$, $\theta = v$, and $w = (\mathrm{e}^{c_2} z_1 z_2 z_3^2, \mathrm{e}^{c_3} z_1 z_3, \mathrm{e}^{c_4} z_1, \mathrm{e}^{c_5} z_2)$, so we get

$$R_1^{\eta_5(3)}(z) = \frac{-(\mathrm{e}^{c_2} z_1 z_2 z_3^2)^{-1}}{1-\mathrm{e}^{-c_2+2c_3-c_4+c_5}}, \quad R_2^{\eta_5(3)}(z) = \frac{(\mathrm{e}^{c_2} z_1 z_2 z_3^2)^{-1}(1+\mathrm{e}^{c_3} z_1 z_3)}{1-\mathrm{e}^{-c_2+2c_3-c_4+c_5}},$$

and

$$R_3^{\eta_5(3)}(z) = \frac{-(\mathrm{e}^{c_4} z_1)^{-1}(\mathrm{e}^{2c_3-c_2} z_1 z_2^{-1})}{1-\mathrm{e}^{-c_2+2c_3-c_4+c_5}}, \quad R_4^{\eta_5(3)}(z) = \frac{\mathrm{e}^{2c_3-c_2-c_4} z_2^{-1}}{1-\mathrm{e}^{-c_2+2c_3-c_4+c_5}}.$$

Notice that the terms associated with $R_1^{\eta_5(3)}(z)$, $R_2^{\eta_5(3)}(z)$, $R_3^{\eta_5(3)}(z)$, and $R_4^{\eta_5(3)}(z)$ *annihilate* the respective entries 2, 3, 4, and 5 in the base $\eta_5(3)$.

Therefore, we have the following expansion:

$$\begin{aligned}
\widehat{F}_d(z,c) &= \frac{Q_{\eta_5(1)}(z) R_1^{\eta_5(1)}(z)}{(1-\mathrm{e}^{c_2} z_1 z_2 z_3^2)(1-\mathrm{e}^{c_4} z_1)(1-\mathrm{e}^{c_5} z_2)(1-\mathrm{e}^{c_6} z_3)} \\
&+ \frac{Q_{\eta_5(1)}(z) R_2^{\eta_5(1)}(z) + Q_{\eta_5(2)}(z) R_2^{\eta_5(2)}(z)}{(1-\mathrm{e}^{c_1} z_1 z_2^2)(1-\mathrm{e}^{c_2} z_1 z_2 z_3^2)(1-\mathrm{e}^{c_5} z_2)(1-\mathrm{e}^{c_6} z_3)} \\
&+ \frac{Q_{\eta_5(1)}(z) R_3^{\eta_5(1)}(z)}{(1-\mathrm{e}^{c_1} z_1 z_2^2)(1-\mathrm{e}^{c_2} z_1 z_2 z_3^2)(1-\mathrm{e}^{c_4} z_1)(1-\mathrm{e}^{c_6} z_3)} \\
&+ \frac{Q_{\eta_5(2)}(z) R_1^{\eta_5(2)}(z)}{(1-\mathrm{e}^{c_1} z_1 z_2^2)(1-\mathrm{e}^{c_3} z_1 z_3)(1-\mathrm{e}^{c_5} z_2)(1-\mathrm{e}^{c_6} z_3)} \\
&+ \frac{Q_{\eta_5(2)}(z) R_3^{\eta_5(2)}(z)}{(1-\mathrm{e}^{c_1} z_1 z_2^2)(1-\mathrm{e}^{c_2} z_1 z_2 z_3^2)(1-\mathrm{e}^{c_3} z_1 z_3)(1-\mathrm{e}^{c_6} z_3)} \\
&+ \frac{Q_{\eta_5(2)}(z) R_4^{\eta_5(2)}(z) + Q_{\eta_5(3)}(z) R_3^{\eta_5(3)}(z)}{(1-\mathrm{e}^{c_1} z_1 z_2^2)(1-\mathrm{e}^{c_2} z_1 z_2 z_3^2)(1-\mathrm{e}^{c_3} z_1 z_3)(1-\mathrm{e}^{c_5} z_2)} \\
&+ \frac{Q_{\eta_5(3)}(z) R_1^{\eta_5(3)}(z)}{(1-\mathrm{e}^{c_1} z_1 z_2^2)(1-\mathrm{e}^{c_3} z_1 z_3)(1-\mathrm{e}^{c_4} z_1)(1-\mathrm{e}^{c_5} z_2)} \\
&+ \frac{Q_{\eta_5(3)}(z) R_2^{\eta_5(3)}(z)}{(1-\mathrm{e}^{c_1} z_1 z_2^2)(1-\mathrm{e}^{c_2} z_1 z_2 z_3^2)(1-\mathrm{e}^{c_4} z_1)(1-\mathrm{e}^{c_5} z_2)} \\
&+ \frac{Q_{\eta_5(3)}(z) R_4^{\eta_5(3)}(z)}{(1-\mathrm{e}^{c_1} z_1 z_2^2)(1-\mathrm{e}^{c_2} z_1 z_2 z_3^2)(1-\mathrm{e}^{c_3} z_1 z_3)(1-\mathrm{e}^{c_4} z_1)} \\
&= \sum_{j=1}^{9} \frac{Q_{\eta_4(j)}(z)}{\prod_{k\in\eta_4(j)}(1-\mathrm{e}^{c_k} z^{A_k})}.
\end{aligned}$$

Final Step: At this last step we obtain the required decomposition (4.50), that is, we will be able to express $\widehat{F}_d(z,c)$ as the sum

$$\widehat{F}_d(z,c) = \sum_j \frac{Q_{\eta_3(j)}(z)}{\prod_{k\in\eta_3(j)}(1-e^{c_k}z^{A_k})}. \tag{4.52}$$

The exact values of $\widehat{f}_d(y,c)$ can then be calculated by direct application of Theorem 4.2. Moreover, we must make the observation that, out of the potentially $\binom{6}{3} = 20$ terms, the above sum contains 16 terms. We next conclude this section by providing the term $Q_{\eta_3(j)}(z)$ relative to the basis $\eta_3(j) = \{2,5,6\} \in \mathbf{J}_3$.

Setting $\eta_4(1) = \{2,4,5,6\} \in \mathbf{J}_4$, choose $\breve{\sigma} := \{4,5,6\}$ and $\sigma := \{2,4,5,6\}$. Let $v := (-1,1,1,2) \in \mathbb{Z}^4$ and solve $A_\sigma v = 0$. Then $q = 4$, $\theta = v$, $w = (e^{c_2}z_1z_2z_3^2, e^{c_4}z_1, e^{c_5}z_2, e^{c_6}z_3)$, and so we get

$$R_1^{\eta_4(1)}(z) = \frac{-(e^{c_2}z_1z_2z_3^2)^{-1}}{1-e^{2c_6+c_5+c_4-c_2}}, \quad R_2^{\eta_4(1)}(z) = \frac{(e^{c_2}z_1z_2z_3^2)^{-1}}{1-e^{2c_6+c_5+c_4-c_2}},$$

and

$$R_3^{\eta_4(1)}(z) = \frac{e^{c_4-c_2}(z_2z_3^2)^{-1}}{1-e^{2c_6+c_5+c_4-c_2}}, \quad R_4^{\eta_4(1)}(z) = \frac{(e^{c_4+c_5-c_2}z_3^{-2})(1+e^{c_6}z_3)}{1-e^{2c_6+c_5+c_4-c_2}}.$$

Notice that the term associated with $R_2^{\eta_4(1)}$ *annihilates* the entry 4 in the base $\eta_4(1) = \{2,4,5,6\}$, so we are getting the desired base $\eta_3(1) = \{2,5,6\}$.

Setting $\eta_4(2) = \{1,2,5,6\} \in \mathbf{J}_4$, choose $\breve{\sigma} := \{2,5,6\}$ and $\sigma := \{1,2,5,6\}$. Let $v := (-1,1,1,-2) \in \mathbb{Z}^4$ and solve $A_\sigma v = 0$. Then $q = 4$, $\theta = v$, $w = (e^{c_1}z_1z_2^2, e^{c_2}z_1z_2z_3^2, e^{c_5}z_2, e^{c_6}z_3)$, and so we get

$$R_1^{\eta_4(2)}(z) = \frac{-(e^{c_1}z_1z_2^2)^{-1}}{1-e^{c_2+c_5-c_1-2c_6}}, \quad R_2^{\eta_4(2)}(z) = \frac{(e^{c_1}z_1z_2^2)^{-1}}{1-e^{c_2+c_5-c_1-2c_6}},$$

and

$$\begin{aligned} R_3^{\eta_4(2)}(z) &= \frac{e^{c_2-c_1}z_2^{-1}z_3^2}{1-e^{c_2+c_5-c_1-2c_6}}, \\ R_4^{\eta_4(2)}(z) &= \frac{-(e^{c_6}z_3)^{-1}(e^{c_2-c_1+c_5}z_3^2)(1+(e^{c_6}z_3)^{-1})}{1-e^{c_2+c_5-c_1-2c_6}}. \end{aligned}$$

Notice that the term associated with $R_1^{\eta_4(2)}$ *annihilates* the entry 1 in the base $\eta_4(2) = \{1,2,5,6\}$, so we are getting the desired base $\eta_3(1) = \{2,5,6\}$.

Therefore, working on the base $\eta_3(1)$, we obtain the numerator

$$\begin{aligned} Q_{\eta_3(1)}(z) &= Q_{\eta_4(1)}R_2^{\eta_4(1)} + Q_{\eta_4(2)}R_1^{\eta_4(2)} \\ &= \left[Q_{\eta_5(1)}(z)R_1^{\eta_5(1)}(z)\right]R_2^{\eta_4(1)} + \left[Q_{\eta_5(1)}(z)R_2^{\eta_5(1)}(z) + Q_{\eta_5(2)}(z)R_2^{\eta_5(2)}(z)\right]R_1^{\eta_4(2)}. \end{aligned}$$

4.6 A simple explicit formula

In this section we are back to primal approaches that work in the space $\mathbb{R}^n$ of primal variables with $y \in \mathbb{Z}^m$ fixed. We provide an explicit expression of $\widehat{f}_d(y,c)$ and an algorithm that involves only simple (possibly numerous) elementary operations. Again, in contrast to Barvinok's algorithm, the computational complexity of this primal approach is not polynomial in the input size for fixed dimension. It uses Brion's formal series formula (4.5) along with an explicit description of the supporting cones at the vertices of $\Omega(y)$. It also has a simple equivalent formulation as a (finite) *group problem*. Finally, we exhibit *finitely many* fixed convex *cones* of $\mathbb{R}^n$ explicitly and exclusively defined from the matrix A, such that for *any* $y \in \mathbb{Z}^m$, $\widehat{f_d}(y,c)$ is obtained by a simple formula involving the evaluation of $\sum_x e^{c'x}$ where the summation is over the integral points of those cones only.

With same notation as in Section 4.3, and $\sigma \in \mathscr{B}(\Delta,\gamma)$, let $\mu_\sigma := \det A_\sigma$, or $|A_\sigma|$ $\mathbb{Z}_\sigma := \{1,\ldots,\mu_\sigma\}$, $A_{\not\sigma} := [A_j]_{j\notin\sigma}$, and let $\delta_\sigma : \mathbb{Z}^m \to \{0,1\}$ be the function

$$z \mapsto \delta_\sigma(z) := \begin{cases} 1 & \text{if } A_\sigma^{-1} z \in \mathbb{Z}^m \\ 0 & \text{otherwise,} \end{cases}$$

$z \in \mathbb{Z}^m$. In what follows one assumes that the convex cone $\{u \in \mathbb{R}^m : A'u > 0\}$ is nonempty so that $\Omega(y)$ is a polytope for every $y \in \mathbb{Z}^m$.

Theorem 4.4. *Let $c \in \mathbb{R}^n$ be regular and let $y \in \gamma \cap \mathbb{Z}^m$ for some chamber γ. Then*

$$\widehat{f_d}(y,c) = \sum_{\sigma \in \mathscr{B}(\Delta,\gamma)} \frac{R_1(y,\sigma;c)}{R_2(\sigma;c)}, \quad \textit{with} \tag{4.53}$$

$$R_1(y,\sigma;c) := e^{c'x(\sigma)} \sum_{u \in \mathbb{Z}_\sigma^{n-m}} \delta_\sigma(y - A_{\not\sigma} u)\, e^{u'(c_{\not\sigma} - \pi_\sigma A_{\not\sigma})} \tag{4.54}$$

$$\textit{and} \quad R_2(\sigma;c) := \prod_{k \notin \sigma} \left[1 - \left(e^{c_k - \pi_\sigma A_k}\right)^{\mu_\sigma}\right]. \tag{4.55}$$

The proof uses Brion's formula (4.5) and an explicit description of the supporting cone $\mathrm{co}(\Omega(y),\hat{x})$ of $\Omega(y)$ at a vertex $\hat{x}$. Indeed, when $y \in \gamma$ then $\Omega(y)$ is simple and at a vertex $x(\sigma)$ with associated basis $\sigma \in \mathscr{B}(\Delta,\gamma)$:

$$\mathrm{co}(\Omega(y),x(\sigma)) = \{(x_\sigma, x_{\not\sigma}) \in \mathbb{R}^n : A_\sigma x_\sigma + A_{\not\sigma} x_{\not\sigma} = y, \quad x_{\not\sigma} \geq 0\}.$$

Even if easy, computing $R_1(\cdot)$ in (4.54) may become quite expensive when μ_σ is large, as one must evaluate μ_σ^{n-m} terms (the cardinality of $\mathbb{Z}_\sigma^{n-m}$). However, as detailed below, a more careful analysis of (4.54) yields some simplifications.

Simplification via group theory

For every basis $\sigma \in \mathscr{B}(\Delta,\gamma)$, consider the lattice $\Lambda_\sigma := \bigoplus_{j\in\sigma} A_j\mathbb{Z} \subset \mathbb{Z}^m$ generated by the columns $(A_j)_{j\in\sigma}$ of A. The finite group

$$\mathscr{G}_\sigma := \mathbb{Z}^m/\Lambda_\sigma = \{\mathrm{Ec}[0,\sigma], \mathrm{Ec}[2,\sigma], \ldots, \mathrm{Ec}[\mu_\sigma - 1,\sigma]\}$$

is commutative with $\mu_\sigma = \det A_\sigma$ or $|A_\sigma|$ equivalence classes $\mathrm{Ec}[j,\sigma]$. Therefore, $\mathscr{G}_\sigma$ is isomorphic to a finite Cartesian product of cyclic groups $\mathbb{Z}_{\eta_k}$, i.e.,

$$\mathscr{G}_\sigma \cong \mathbb{Z}_{\eta_1} \times \mathbb{Z}_{\eta_2} \times \cdots \times \mathbb{Z}_{\eta_s}$$

with $\mu_\sigma = \eta_1 \eta_2 \ldots \eta_s$. (When $\mu_\sigma = 1$ then $\mathscr{G}_\sigma$ is the trivial group $\{0\}$.) Notice that the Cartesian product $\mathbb{Z}_{\eta_1} \times \cdots \times \mathbb{Z}_{\eta_s}$ can be seen as $\mathbb{Z}^s$ modulo the vector $\eta := (\eta_1, \eta_2, \ldots, \eta_s)' \in \mathbb{N}^s$. Hence, for every finite commutative group $\mathscr{G}_\sigma$, there exists a positive integer $s_\sigma \geq 1$, a vector $\eta_\sigma \in \mathbb{N}^{s_\sigma}$ with positive entries, and a group isomorphism,

$$g_\sigma : \mathscr{G}_\sigma \rightarrow \mathbb{Z}^{s_\sigma} \bmod \eta_\sigma,$$

where $g_\sigma(\xi) \bmod \eta_\sigma$ means evaluating $[g_\sigma(\xi)]_k \bmod [\eta_\sigma]_k$ for every $k = 1, \ldots, s_\sigma$. For every $y \in \mathbb{Z}^m$ there exists a unique equivalence class $\mathrm{Ec}[j(y),\sigma] \ni y$ that we use to define the group epimorphism, $\hat{h}_\sigma : \mathbb{Z}^m \rightarrow \mathbb{Z}^{s_\sigma} \bmod \eta_\sigma$, by

$$y \mapsto \hat{h}_\sigma(y) := g_\sigma(\mathrm{Ec}[j(y),\sigma]).$$

The unit element of $\mathscr{G}_\sigma$ is the equivalence class $\mathrm{Ec}[0,\sigma] = \{A_\sigma q \mid q \in \mathbb{Z}^m\}$. Hence $\hat{h}_\sigma(y) = 0$ iff $y = A_\sigma q$ for some $q \in \mathbb{Z}^m$. So define $\delta_\sigma : \mathbb{Z}^m \rightarrow \{0,1\}$ by

$$y \mapsto \delta_\sigma(y) := \begin{cases} 1 & \text{if } \hat{h}_\sigma(y) = 0, \\ 0 & \text{otherwise.} \end{cases}$$

Next, given any matrix $B = [B_1|\cdots|B_t] \in \mathbb{Z}^{m \times t}$, let

$$\hat{h}_\sigma(B) := [\hat{h}_\sigma(B_1)|\hat{h}_\sigma(B_2)|\cdots|\hat{h}_\sigma(B_t)] \in \mathbb{Z}^{s_\sigma \times t},$$

so that $\hat{h}_\sigma(y - A_\phi u) \equiv \hat{h}_\sigma(y) - \hat{h}_\sigma(A_\phi)u \bmod \eta_\sigma$.

Given $q \in \mathbb{Z}^m$, $\mu_\sigma A_\sigma^{-1} q \in \mathbb{Z}^m$ but μ_σ may not be the smallest positive integer with that property. Therefore, given $\sigma \in \mathscr{B}(\Delta,\gamma)$ and $k \notin \sigma$, define $\nu_{k,\sigma} \geq 1$ to be *order* of $\hat{h}_\sigma(A_k)$. That is, $\nu_{k,\sigma}$ is the smallest positive integer such that $\nu_{k,\sigma}\hat{h}_\sigma(A_k)$ vanishes modulus η_σ, or equivalently $\nu_{k,\sigma} A_\sigma^{-1} A_k \in \mathbb{Z}^m$. Let

$$R_2^*(\sigma;c) := \prod_{k \notin \sigma} \left[1 - \left(e^{c_k - \pi_\sigma A_k}\right)^{\nu_{k,\sigma}}\right] \tag{4.56}$$

and

$$\begin{aligned} R_1^*(y,\sigma;c) &:= e^{c'x(\sigma)} \sum_{u \in U_\phi} \delta_\sigma(y - A_\phi u)\, e^{u'(c_\phi - \pi_\sigma A_\phi)} \\ &= e^{c'x(\sigma)} \sum_{u \in U_\phi} \left\{ e^{u'(c_\phi - \pi_\sigma A_\phi)} : \hat{h}_\sigma(y) \equiv \hat{h}_\sigma(A_\phi)u \bmod \eta_\sigma \right\}, \end{aligned} \tag{4.57}$$

with $U_\phi \subset \mathbb{Z}^{n-m}$ being the set $\{u \in \mathbb{Z}^{n-m} \,|\, 0 \leq u_k \leq \nu_{k,\sigma} - 1\}$.

Corollary 4.2. *Let $c \in \mathbb{R}^n$ be regular and let $y \in \gamma \cap \mathbb{Z}^m$ for some chamber γ. Then*

$$\widehat{f}_d(y,c) = \sum_{\sigma \in \mathscr{B}(\Delta,\gamma)} \frac{R_1^*(y,\sigma;c)}{R_2^*(\sigma;c)}, \tag{4.58}$$

with R_1^*, R_2^* as in (4.56) and (4.57).

So in Corollary 4.2 the evaluation of R_1 in Theorem 4.4 is replaced with the simpler finite group problem of evaluating R_1^* in (4.57).

One may even proceed further. Pick up a *minimal* representative element in every class $\mathrm{Ec}[j,\sigma]$, i.e., fix

$$\xi_j^\sigma \in \mathrm{Ec}[j,\sigma] \quad \text{such that} \quad A_\sigma^{-1} y \geq A_\sigma^{-1} \xi_j^\sigma \geq 0 \tag{4.59}$$

for every $y \in \mathrm{Ec}[j,\sigma]$ with $A_\sigma^{-1} y \geq 0$. The minimal element ξ_j^σ is computed as follows: Let $d \in \mathrm{Ec}[j,\sigma]$ be arbitrary and let $d^* \in \mathbb{Z}^m$ be such that $d_k^* := \lceil -(A_\sigma^{-1} d)_k \rceil$, $k = 1,\ldots,m$. Then $\xi_j^\sigma := d + A_\sigma d^*$ satisfies (4.59) and $A_\sigma^{-1} y \geq 0$ holds iff

$$y = \xi_j^\sigma + A_\sigma q \quad \text{with } q \in \mathbb{N}^m. \tag{4.60}$$

Theorem 4.5. *Let $c \in \mathbb{R}^n$ be regular, $y \in \mathbb{Z}^m \cap \gamma$ for some given chamber γ, and let $\xi_{j,\sigma}$ be as in (4.59). For every basis $\sigma \in \mathscr{B}(\Delta,\gamma)$ there is a unique index $0 \leq j(y) < \mu_\sigma$ such that $y \in \mathrm{Ec}[j(y),\sigma]$, and*

$$\widehat{f}_d(y,c) = \sum_{\sigma \in \mathscr{B}(\Delta,\gamma)} \frac{R_1(\xi_{j(y)}^\sigma,\sigma;c)}{R_2(\sigma;c)} \, e^{c_\sigma' \lfloor A_\sigma^{-1} y \rfloor}, \tag{4.61}$$

with the convention $\lfloor A_\sigma{}^{-1} y \rfloor = (\lfloor (A_\sigma{}^{-1} y)_k \rfloor) \in \mathbb{Z}^m$.

Theorem 4.5 states that in order to compute $\widehat{f}_d(y,c)$ for arbitrary values $y \in \mathbb{Z}^m \cap \gamma$, it suffices to compute $R_1(v,\sigma;c)$ (or $R_1^*(v,\sigma;c)$) for *finitely* many explicit values $v = \xi_j^\sigma$, with $\sigma \in \mathscr{B}(\Delta,\gamma)$ and $0 \leq j < \mu_\sigma$, and one obtains $\widehat{f}_d(y,c)$ via (4.61). In other words, in a chamber γ, one only needs to consider *finitely many* fixed convex cones $C(\Omega(\xi_j^\sigma),\sigma) \subset \mathbb{R}^n$, where $\sigma \in \mathscr{B}(\Delta,\gamma)$ and $0 \leq j < \mu_\sigma$, and compute their associated rational function (4.61). (For the whole space $\mathbb{Z}^m$ it suffices to consider *all* chambers γ and all cones $C(\Omega(\xi_{j,\sigma}),\sigma) \subset \mathbb{R}^n$, where $\sigma \in \mathscr{B}(\Delta,\gamma)$ and $0 \leq j < \mu_\sigma$.) The counting function $\widehat{f}_d(y,c)$ is then obtained as follows.

Algorithm 2 to compute $\widehat{f}_d(y,c)$:

Input: $y \in \mathbb{Z}^m \cap \gamma, c \in \mathbb{R}^n$.
Output: $\rho = \widehat{f}_d(y,c)$.
Set $\rho := 0$. For every $\sigma \in \mathscr{B}(\Delta,\gamma)$

- compute $\xi^{\sigma}_{j(y)} := y - A_\sigma \lfloor A_\sigma^{-1} y \rfloor \in \mathbb{Z}^m$.
- read the value $R_1(\xi^{\sigma}_{j(y)}, \sigma; c)/R_2(\sigma, c)$, and update ρ by

$$\rho := \rho + \frac{R_1(\xi^{\sigma}_{j(y)} \sigma; c)}{R_2(\sigma; c)} \mathrm{e}^{c'_\sigma \lfloor A_\sigma^{-1} y \rfloor}.$$

Finally, in view of (4.54)–(4.55), the above algorithm can be symbolic, i.e., $w := \mathrm{e}^c \in \mathbb{R}^n$ can be treated symbolically, and $\rho = \widehat{f}_d(y,w)$ becomes a rational fraction of w.

4.7 Notes

Most of the material is from Barvinok [14], Barvinok and Pommersheim [15], Barvinok and Woods [16], Brion and Vergne [27], and Lasserre and Zeron [98, 99, 100]. The Markov chain Monte Carlo method for approximate counting and its complexity are nicely discussed in, e.g., Jerrum and Sinclair [72].

Another algebraic approach that avoids computing residues was also proposed in [98]. It is based on Hilbert's Nullstellensatz and as in Section 4.5, it provides a special decomposition of $\widehat{F}_d(z,c)$ into certain partial fractions. Again, the main work is the computation of the coefficients of the polynomials involved in this decomposition, which reduces to solving a linear system as bounds on the degree of the polynomials in this decomposition are available (see, e.g., Seidenberg [122] and Kollár [82]). However, this approach works only for relatively small values of m. Beck [20] also uses a partial fraction decomposition approach for counting lattice points in a polytope.

There are also specialized algorithms when the matrix A is *unimodular*, in which case Brion and Vergne's formula (4.14) simplifies significantly. See for instance the works of Baldoni-Silva and Vergne [11], Baldoni et al. [12], Beck [17], and Beck Diaz, and Robins [19], as well as the software developed in Cochet's thesis [33] with several applications, most notably in algebraic geometry. The unimodular case is also considered in De Loera and Sturmfels [44] using a Gröbner base approach as well as Barvinok's algorithm.

The primal approach that uses Barvinok's algorithm has been implemented in the software package `LattE` developed by De Loera et al. [40, 41], available at `http://www.math.ucdavis.edu/~latte/`.

Finally, for details on approximate counting via *Markov chain Monte Carlo* methods, the interested reader is referred to, e.g., Dyer et al. [50, 48] and Jerrum and Sinclair [72].

Chapter 5
Relating the Discrete Problems $\mathbf{P}_d$ and $\mathbf{I}_d$ with $\mathbf{P}$

5.1 Introduction

In the respective *discrete* analogues $\mathbf{P}_d$ and $\mathbf{I}_d$ of (3.1) and (3.2), one replaces the positive cone in $\mathbb{R}^n$ ($x \geq 0$) by $\mathbb{N}^n$, that is, with $A \in \mathbb{Z}^{m\times n}$, $b \in \mathbb{Z}^m$, and $c \in \mathbb{R}^m$, (3.1) becomes the integer program

$$\mathbf{P}_d : f_d(y,c) := \max_x \{c'x \,|\, Ax = y, \quad x \in \mathbb{N}^n\}, \tag{5.1}$$

whereas (3.2) becomes a summation over $\mathbb{N}^n \cap \Omega(y)$, i.e.,

$$\mathbf{I}_d : \widehat{f}_d(y,c) := \left\{ \sum_x \mathrm{e}^{c'x} \,|\, Ax = y, \quad x \in \mathbb{N}^n \right\}. \tag{5.2}$$

In Chapter 4 we proposed an algorithm to compute $\widehat{f}_d(y,c)$. In this chapter we see how $\mathbf{I}_d$ and its *dual* $\mathbf{I}_d^*$ provide insights into the discrete optimization problem $\mathbf{P}_d$.

We first compare $\mathbf{I}^*$ and $\mathbf{I}_d^*$, which are of the same nature, that is, complex integrals that one can solve by Cauchy's residue techniques as in Chapter 2 and 4, respectively. However, there are important differences between $\mathbf{I}^*$ and $\mathbf{I}_d^*$ that make the latter problem much more difficult to solve. We next make a comparison between $\mathbf{I}_d$ and $\mathbf{P}_d$ pretty much in the spirit of the comparison that we already made between $\mathbf{I}$, $\mathbf{I}^*$ and $\mathbf{P}$, $\mathbf{P}^*$ in Section 3.2. The insights provided by $\mathbf{I}_d^*$ permit us to define precisely what are the *dual* variables of $\mathbf{P}_d$ and how they are related to the well-known dual variables of $\mathbf{P}^*$.

Again, the approach we follow is *dual* to the algebraic methods described in [126, 128] as well as in [58], where integer programming appears as an *arithmetic refinement* of linear programming. In contrast, we work in the range space ($\subset \mathbb{Z}^m$) of the linear mapping A instead of the primal space

J.B. Lasserre, *Linear and Integer Programming vs Linear Integration and Counting*,
Springer Series in Operations Research and Financial Engineering, DOI 10.1007/978-0-387-09414-4_5,

$\mathbb{Z}^n$. The variables of interest associated with the constraints $Ax = y$ are the analogues of the usual *dual* variables in linear programming, except they are now in $\mathbb{C}^m$ rather than in $\mathbb{R}^m$.

5.2 Comparing the dual problems $\mathbf{I}^*$ and $\mathbf{I}_d^*$

We saw in Chapter 3 that the dual problem $\mathbf{I}^*$ (3.8) is the inversion process at the point $y \in \mathbb{R}^m$ of the Laplace transform $\widehat{F}(\cdot,c)$ of $\widehat{f}(\cdot,c)$, exactly as the dual problem $\mathbf{P}^*$ is the inversion process at $y \in \mathbb{R}^m$ of the Legendre–Fenchel transform $F(\cdot,c)$ of $f(\cdot,c)$. In the former the inversion process is the inverse Laplace transform, whereas in the latter it is again the Legendre–Fenchel transform, which is involutive for convex functions.

Similarly, the dual problem $\mathbf{I}_d^*$ in (2.3) is the inversion process at the point $y \in \mathbb{R}^m$, of the $\mathbb{Z}$-transform (or generating function) $\widehat{F_d}(\cdot,c)$ of $\widehat{f_d}(\cdot,c)$, i.e., the perfect discrete analogue of $\mathbf{I}^*$, as the $\mathbb{Z}$-transform is the perfect analogue of the Laplace transform. One has the parallel between the duality types shown in the following table:

Table 5.1 Parallel duality, ...

Continuous Laplace Duality	**Discrete $\mathbb{Z}$-Duality**
$\widehat{f}(y,c) := \int_{Ax=y;\,x\in\mathbb{R}^n_+} e^{c'x}\,d\sigma$	$\widehat{f_d}(y,c) := \sum_{Ax=y;\,x\in\mathbb{N}^n} e^{c'x}$
$\widehat{F}(\lambda,c) := \int_{\mathbb{R}^m} e^{-\lambda' y}\widehat{f}(y,c)dy = \prod_{k=1}^{n} \frac{1}{(A'\lambda - c)_k}$	$\widehat{F_d}(z,c) := \sum_{y\in\mathbb{Z}^m} z^{-y}\widehat{f_d}(y,c) = \prod_{k=1}^{n} \frac{1}{1-e^{c_k}z^{-A_k}}$
with $\Re(A'\lambda - c) > 0$.	with $\lvert z^A\rvert > e^c$
$\mathbf{I}^* : \int_{\gamma\pm i\infty} e^{y-e_m}\widehat{F}(\lambda,c)\,d\lambda$	$\mathbf{I}_d^* : \int_{\lvert z\rvert=\rho} e^{y-e_m}\widehat{F_d}(z,c)\,dz$
with $A'\gamma > c$	with $A'\ln\rho > c$

Observe that both dual problems $\mathbf{I}^*$ and $\mathbf{I}_d^*$ are of same nature as both reduce to computation of a complex integral whose integrand is a rational function. In particular, we saw in Chapter 2 and 4, respectively, how both problems $\mathbf{I}^*$ and $\mathbf{I}_d^*$ can be solved by Cauchy residue techniques. Notice also that the respective domains of definition of $\widehat{F}$ and $\widehat{F_d}$ are identical. Indeed, writing $z = e^{\lambda}$ with $\lambda \in \mathbb{C}$ yields $\lvert z^A\rvert \geq 1 \Leftrightarrow \Re(A'\lambda - c) \geq 0$. However, a quick look at the residue algorithms (Section 2.4 for $\mathbf{I}^*$ and Section 4.4 for $\mathbf{I}_d^*$) reveals immediately that $\mathbf{I}_d^*$ is much more difficult to solve than $\mathbf{I}^*$, despite that both integrands, the Laplace transform $\widehat{F}$ and the $\mathbb{Z}$-transform $\widehat{F_d}$, share some similarities. Indeed, there is an important difference between $\mathbf{I}^*$ and $\mathbf{I}_d^*$. While in $\mathbf{I}^*$ the data

$\{A_{jk}\}$ appear as *coefficients* of the dual variables λ_k in $\widehat{F}(\lambda,c)$, in $\mathbf{I}_d^*$ they now appear as *exponents* of the dual variables z_k in $\widehat{F_d}(z,c)$. As an immediate consequence, the rational function $\widehat{F_d}(\cdot,c)$ has many more poles than $\widehat{F}(\cdot,c)$ (by considering one variable at a time), and in particular, many of them are *complex*, whereas $\widehat{F}(\cdot,c)$ has only *real* poles. As a result, the integration of $\widehat{F_d}(z,c)$ is more complicated than that of $\widehat{F}(\lambda,c)$. This is reflected not only in the residue algorithms of Section 2.4 and Section 4.4, but also in the respective Brion and Vergne's continuous and discrete formulas (2.15) and (4.14). However, we will see that the poles of $\widehat{F_d}(z,c)$ are still simply related to those of $\widehat{F}(\lambda,c)$.

5.3 A dual comparison of P and $\mathbf{P}_d$

We are now in a position to see how $\mathbf{I}_d^*$ provides some nice information about the optimal value $f_d(b,c)$ of the discrete optimization problem $\mathbf{P}_d$.

Recall that $A\in\mathbb{Z}^{m\times n}$ and $y\in\mathbb{Z}^m$, which implies in particular that the *lattice* $\Lambda:=A(\mathbb{Z}^n)$ is a sublattice of $\mathbb{Z}^m$ ($\Lambda\subset\mathbb{Z}^m$). Note that y in (5.1) and (5.2) is necessarily in Λ. Recall also the notation used in Section 2.3 and Section 4.3.

Theorem 5.1. *Let $A\in\mathbb{Z}^{m\times n}$, $y\in\mathbb{Z}^m$ and let $c\in\mathbb{Z}^n$ be regular with $-c$ in the interior of $(\mathbb{R}^n_+\cap V)^*$. Let $y\in\overline{\gamma}\cap A(\mathbb{Z}^n)$ and let $q\in\mathbb{N}$ be the least common multiple (l.c.m.) of $\{\mu(\sigma)\}_{\sigma\in\mathscr{B}(\Delta,\gamma)}$.*

If $Ax=y$ has no solution $x\in\mathbb{N}^n$ then $f_d(y,c)=-\infty$. Otherwise, if

$$\max_{x(\sigma):\,\sigma\in\mathscr{B}(\Delta,\gamma)}\left[c'x(\sigma)+\lim_{r\to\infty}\frac{1}{r}\ln U_\sigma(y,rc)\right] \tag{5.3}$$

is attained at a unique vertex $x(\sigma)$ of $\Omega(y)$, then

$$\begin{aligned} f_d(y,c) &= \max_{x(\sigma):\,\sigma\in\mathscr{B}(\Delta,\gamma)}\left[c'x(\sigma)+\lim_{r\to\infty}\frac{1}{r}\ln U_\sigma(y,rc)\right]\\ &= \max_{x(\sigma):\,\sigma\in\mathscr{B}(\Delta,\gamma)}\left[c'x(\sigma)+\frac{1}{q}(\deg P_{\sigma y}-\deg Q_{\sigma y})\right] \end{aligned} \tag{5.4}$$

for some real-valued univariate polynomials $P_{\sigma y}$, $Q_{\sigma y}$.

Moreover, the term $\lim_{r\to\infty}\ln U_\sigma(y,rc)/r$ or $(\deg P_{\sigma y}-\deg Q_{\sigma y})/q$ in (5.4) is a sum of certain reduced costs $c_k-\pi_\sigma A_k$ (with $k\notin\sigma$).

A proof is postponed until Section 5.4. Of course, (5.4) is not easy to evaluate but it shows that the optimal value $f_d(y,c)$ of $\mathbf{P}_d$ is strongly related to the various complex poles of $\widehat{F_d}(z,c)$. It is also interesting to note the crucial role played by the reduced costs $(c_k-\pi_\sigma A_k)$ in linear programming. Indeed, from the proof of Theorem 5.1, the optimal value $f_d(y,c)$ is the value of $c'x$ at some vertex $x(\sigma)$ plus a sum of certain reduced costs. Thus, as for the linear program **P**, the optimal value $f_d(y,c)$ of $\mathbf{P}_d$ can be found by inspection of (certain sums of) reduced costs associated with each vertex of

$\Omega(y)$. In fact, when the max in (5.4) is attained at a unique vertex, this vertex is the optimal vertex σ^* of the linear program $\mathbf{P}$. We come back to this in more detail in Chapter 6.

We next derive an asymptotic result that relates the respective optimal values $f_d(y,c)$ and $f(y,c)$ of $\mathbf{P}_d$ and $\mathbf{P}$.

Corollary 5.1. *Let $A \in \mathbb{Z}^{m\times n}$, $y \in \mathbb{Z}^m$ and let $c \in \mathbb{R}^n$ be regular with $-c$ in the interior of $(\mathbb{R}^n_+ \cap V)^*$. Let $y \in \gamma \cap \Lambda$ and let $x^* \in \Omega(y)$ be an optimal vertex of $\mathbf{P}$, that is, $f(y,c) = c'x^* = c'x(\sigma^*)$ for $\sigma^* \in \mathscr{B}(\Delta,\gamma)$, the unique optimal basis of $\mathbf{P}$. Then for $t \in \mathbb{N}$ sufficiently large,*

$$f_d(ty,c) - f(ty,c) = \lim_{r\to\infty}\left[\frac{1}{r}\ln U_{\sigma^*}(ty,rc)\right]. \tag{5.5}$$

In particular, for $t \in \mathbb{N}$ sufficiently large, the function $t \mapsto f(ty,c) - f_d(ty,c)$ is periodic (constant) with period $\mu(\sigma^*)$.

A proof is postponed until Section 5.4. Thus, when $y \in \gamma \cap \Lambda$ is sufficiently large, say $y = ty_0$ with $y_0 \in \Lambda$ and $t \in \mathbb{N}$, the max in (5.3) is attained at the unique optimal basis σ^* of the linear program $\mathbf{P}$ (see details in Section 5.4).

From Theorem 5.1 it also follows that for sufficiently large $t \in \mathbb{N}$, the optimal value $f_d(ty,c)$ is equal to $f(ty,c)$ plus a certain sum of *reduced costs* $(c_k - \pi_{\sigma^*}A_k)$ (with $k \notin \sigma^*$) with respect to the optimal basis σ^*.

We now provide an alternative formulation of Brion and Vergne's discrete formula (4.14), one which explicitly relates dual variables of $\mathbf{P}$ and what we also call dual variables of $\mathbf{P}_d$. Recall that a *feasible basis* of the linear program $\mathbf{P}$ is a basis $\sigma \in \mathscr{B}(\Delta)$ for which $A_\sigma^{-1}y \geq 0$. Thus, let $\sigma \in \mathscr{B}(\Delta)$ be a feasible basis of the linear program $\mathbf{P}$, and consider the system of polynomial equations $z^{A_\sigma} = e^{c_\sigma}$ in $\mathbb{C}^m$ (where $c_\sigma \in \mathbb{R}^m$ is the vector $\{c_j\}_{j\in\sigma}$), i.e.,

$$z_1^{A_{1j}} \cdots z_m^{A_{mj}} = e^{c_j}, \ \forall j \in \sigma. \tag{5.6}$$

The above system (5.6) has $\rho(\sigma)$ $(= \det A_\sigma)$ solutions $\{z(k)\}_{k=1}^{\rho(\sigma)}$ in the form

$$z(k) = e^{\lambda} e^{2i\pi\theta(k)}, \ k = 1,\ldots,\rho(\sigma) \tag{5.7}$$

for $\rho(\sigma)$ vectors $\{\theta(k)\}$ in $\mathbb{R}^m$.

Indeed, writing $z = e^{\lambda}e^{2i\pi\theta}$ (i.e., the vector $\{e^{\lambda_j}e^{2i\pi\theta_j}\}_{j=1}^m$ in $\mathbb{C}^m$), and passing to the logarithm in (5.6), yields

$$A'_\sigma\lambda + 2i\pi A'_\sigma\theta = c_\sigma. \tag{5.8}$$

Thus, $\lambda \in \mathbb{R}^m$ is the unique solution of $A'_\sigma\lambda = c_\sigma$, and θ satisfies

$$A'_\sigma\theta \in \mathbb{Z}^m. \tag{5.9}$$

Equivalently, θ belongs to $(\oplus_{j\in\sigma}A_j\mathbb{Z})^*$, the dual lattice of $\oplus_{j\in\sigma}A_j\mathbb{Z}$.

Thus, there is a one-to-one correspondence between the $\rho(\sigma)$ solutions $\{\theta(k)\}$ and the finite group $G'(\sigma) = (\oplus_{j\in\sigma}A_j\mathbb{Z})^*/\mathbb{Z}^m$. With $s := \mu(\sigma)$, recall that $G(\sigma) = (\oplus_{j\in\sigma}A_j\mathbb{Z})^*/\Lambda^*$ and so $G(\sigma) = \{g_1,\dots,g_s\}$ is a subgroup of $G'(\sigma)$. Define the mapping $\theta : G(\sigma)\to\mathbb{R}^m$,

$$g \mapsto \theta_g := (A'_\sigma)^{-1}g,$$

so that for every character $e^{2i\pi y}$ of $G(\sigma)$, $y \in \Lambda$, we have

$$e^{2i\pi y}(g) = e^{2i\pi y'\theta_g}, \; y \in \Lambda, g \in G(\sigma). \tag{5.10}$$

and

$$e^{2i\pi A_j}(g) = e^{2i\pi A'_j\theta_g} = 1, \; j \in \sigma. \tag{5.11}$$

So, for every $\sigma \in \mathscr{B}(\Delta)$, denote by $\{z_g\}_{g\in G(\sigma)}$ these $\mu(\sigma) \le \rho(\sigma)$ solutions of (5.7), that is,

$$z_g = e^{\lambda}e^{2i\pi\theta_g} \in \mathbb{C}^m, \; g \in G(\sigma), \tag{5.12}$$

with $\lambda = (A'_\sigma)^{-1}c_\sigma$, and where $e^\lambda \in \mathbb{R}^m$ is the vector $\{e^{\lambda_i}\}_{i=1}^m$.

So, in the linear program **P** we have a dual vector $\lambda \in \mathbb{R}^m$ associated with each basis σ. In the integer program $\mathbf{P}_d$, with each (same) basis σ are now associated $\mu(\sigma)$ dual (complex) vectors $\lambda + 2i\pi\theta_g$, $g \in G(\sigma)$. Hence, with a basis σ in linear programming, $\{\lambda + 2i\pi\theta_g\}$ are what we call the *dual variables* of $\mathbf{P}_d$ with respect to the basis σ; they are obtained from (a) the corresponding dual variables $\lambda \in \mathbb{R}^m$ in linear programming and (b) a periodic correction term $2i\pi\theta_g \in \mathbb{C}^m$, $g \in G(\sigma)$.

We next introduce what we call the *vertex residue function*.

Definition 5.1. Let $y \in \Lambda$ and let $c \in \mathbb{R}^n$ be regular. Let $\sigma \in \mathscr{B}(\Delta)$ be a feasible basis of the linear program **P** and for every $r \in \mathbb{N}$, let $\{z_{gr}\}_{g\in G(\sigma)}$ be as in (5.12), with rc in lieu of c, that is,

$$z_{gr} = e^{r\lambda}e^{2i\pi\theta_g} \in \mathbb{C}^m, \; g \in G(\sigma), \quad \text{with } \lambda = (A'_\sigma)^{-1}c_\sigma.$$

The vertex residue function associated with the basis σ of the linear program **P** is the function $R_\sigma(z_g,\cdot) : \mathbb{N}\to\mathbb{R}$ defined by

$$r \mapsto R_\sigma(z_g, r) := \frac{1}{\mu(\sigma)} \sum_{g\in G(\sigma)} \frac{z_{gr}^y}{\prod_{j\notin\sigma}(1 - z_{gr}^{-A_k}e^{rc_k})}, \tag{5.13}$$

which is well-defined because when c is regular, $|z_{gr}|^{A_k} \neq e^{rc_k}$ for all $k \notin \sigma$.

The name *vertex residue* is now clear because in the integration (4.12), $R_\sigma(z_g, r)$ is to be interpreted as a *generalized* Cauchy residue with respect to the $\mu(\sigma)$ poles $\{z_{gr}\}$ of the generating function $\widehat{F}_d(z, rc)$.

Recall that from Corollary 5.1 , when $y \in \gamma\cap\Lambda$ is sufficiently large, say $y = ty_0$ with $y_0 \in \Lambda$ and some large $t \in \mathbb{N}$, the max in (5.4) is attained at the unique optimal basis σ^* of the linear program **P**.

Proposition 5.1. *Let c be regular with $-c \in (\mathbb{R}^n_+ \cap V)^*$, and let $y \in \gamma \cap \Lambda$ be sufficiently large so that the max in (5.3) is attained at the unique optimal basis σ^* of the linear program* $\mathbf{P}$. *Let $\{z_{gr}\}_{g\in G(\sigma^*)}$ be as in Definition 5.1 with $\sigma = \sigma^*$.*

Then the optimal value of $\mathbf{P}_d$ satisfies

$$f_d(y,c) = \lim_{r\to\infty} \frac{1}{r} \ln \left[\frac{1}{\mu(\sigma^*)} \sum_{g\in G(\sigma^*)} \frac{z_{gr}^y}{\prod_{k\notin\sigma^*}(1 - z_{gr}^{-A_k} e^{rc_k})} \right]$$
$$= \lim_{r\to\infty} \frac{1}{r} \ln R_{\sigma^*}(z_g, r), \tag{5.14}$$

and the optimal value of $\mathbf{P}$ *satisfies*

$$f(y,c) = \lim_{r\to\infty} \frac{1}{r} \ln \left[\frac{1}{\mu(\sigma^*)} \sum_{g\in G(\sigma^*)} \frac{|z_{gr}|^y}{\prod_{k\notin\sigma^*}(1 - |z_{gr}|^{-A_k} e^{rc_k})} \right]$$
$$= \lim_{r\to\infty} \frac{1}{r} \ln R_{\sigma^*}(|z_g|, r). \tag{5.15}$$

A proof is postponed until Section 5.4. Proposition 5.1 shows that there is indeed a strong relationship between the integer program $\mathbf{P}_d$ and its continuous analogue, the linear program $\mathbf{P}$. Both optimal values obey exactly the same formula (5.14), but for the continuous version, the complex vector $z_g \in \mathbb{C}^m$ is replaced with the vector $|z_g| = e^{\lambda^*} \in \mathbb{R}^m$ of its component moduli, where $\lambda^* \in \mathbb{R}^m$ is the optimal solution of the LP dual $\mathbf{P}^*$ of $\mathbf{P}$. In summary, when $c \in \mathbb{R}^n$ is regular and $y \in \gamma \cap \Lambda$ is sufficiently large, we have the correspondence displayed in Table 5.2.

Table 5.2 Comparing $\mathbf{P}$ and $\mathbf{P}_d$

Linear Program P	**Integer Program $\mathbf{P}_d$**
Unique optimal basis σ^*	Unique optimal basis σ^*
One optimal dual vector $\lambda^* \in \mathbb{R}^m$	$\mu(\sigma^*)$ dual vectors $z_g \in \mathbb{C}^m$, $g \in G(\sigma^*)$ $\ln z_g = \lambda^* + 2i\pi\theta_g$
$f(y,c) = \lim_{r\to\infty} \frac{1}{r} \ln R_{\sigma^*}(\lvert z_g\rvert, r)$	$f_d(y,c) = \lim_{r\to\infty} \frac{1}{r} \ln R_{\sigma^*}(z_g, r)$

5.4 Proofs

Proof of Theorem 5.1

Use (1.3) and (4.14)–(4.15) to obtain

$$\begin{aligned} e^{f_d(y,c)} &= \lim_{r\to\infty}\left[\sum_{x(\sigma):\,\sigma\in\mathscr{B}(\Delta,\gamma)} \frac{e^{rc'x(\sigma)}}{\mu(\sigma)} U_\sigma(y,rc)\right]^{1/r} \\ &= \lim_{r\to\infty}\left[\sum_{x(\sigma):\,\sigma\in\mathscr{B}(\Delta,\gamma)} H_\sigma(y,rc)\right]^{1/r}. \end{aligned} \tag{5.16}$$

Next, from the expression of $U_\sigma(y,c)$ in (4.15), and with rc in lieu of c, we see that $U_\sigma(y,rc)$ is a function of the variable $u = e^r$, which in turn implies that $H_\sigma(y,rc)$ is also a function of u, of the form

$$H_\sigma(y,rc) = (e^r)^{c'x(\sigma)} \sum_{g\in G(\sigma)} \frac{e^{2i\pi y}(g)}{\sum_j \left(\delta_j(\sigma,g,A)\times (e^r)^{\alpha_j(\sigma,c)}\right)}, \tag{5.17}$$

for finitely many coefficients $\{\delta_j(\sigma,g,A),\alpha_j(\sigma,c)\}$. Note that the coefficients $\alpha_j(\sigma,c)$ are sums of some reduced costs $(c_k - \pi_\sigma A_k)$ (with $k\notin\sigma$). In addition, the (complex) coefficients $\{\delta_j(\sigma,g,A)\}$ do not depend on y.

Let $u := e^{r/q}$ with q be the l.c.m. of $\{\mu(\sigma)\}_{\sigma\in\mathscr{B}(\Delta,\gamma)}$. As $q(c_k-\pi_\sigma A_k)\in\mathbb{Z}$ for all $k\notin\sigma$,

$$H_\sigma(y,rc) = u^{qc'x(\sigma)} \times \frac{P_{\sigma y}(u)}{Q_{\sigma y}(u)} \tag{5.18}$$

for some polynomials $P_{\sigma y}, Q_{\sigma y}\in\mathbb{R}[u]$. In view of (5.17), the degree of $P_{\sigma y}$ and $Q_{\sigma y}$, which depends on y but *not* on the magnitude of y, is uniformly bounded in y. Therefore, as $r\to\infty$,

$$H_\sigma(y,rc) \approx u^{qc'x(\sigma)+\deg P_{\sigma y}-\deg Q_{\sigma y}}, \tag{5.19}$$

so that the limit in (5.16), which is given by $\max_\sigma e^{c'x(\sigma)} \lim_{r\to\infty} U_\sigma(y,rc)^{1/r}$ (as we have assumed unicity of the maximizer σ), is also

$$\max_{x(\sigma):\,\sigma\in\mathscr{B}(\Delta,\gamma)} e^{c'x(\sigma)+(\deg P_{\sigma y}-\deg Q_{\sigma y})/q}.$$

Therefore, $f_d(y,c) = -\infty$ if $Ax = y$ has no solution $x\in\mathbb{N}^n$ and

$$f_d(y,c) = \max_{x(\sigma):\,\sigma\in\mathscr{B}(\Delta,\gamma)} \left[c'x(\sigma) + \frac{1}{q}(\deg P_{\sigma y} - \deg Q_{\sigma y})\right] \tag{5.20}$$

otherwise, from which (5.4) follows easily. □

Proof of Corollary 5.1

Let $t \in \mathbb{N}$ and note that $f(ty,rc) = trf(y,c) = trc'x^* = trc'x(\sigma^*)$. As in the proof of Theorem 5.1, and with ty in lieu of y, we have

$$\widehat{f}_d(ty,rc)^{1/r} = \mathrm{e}^{tc'x^*}\left[\frac{U_{\sigma^*}(ty,rc)}{\mu(\sigma^*)} + \sum_{x(\sigma)\neq x^*:\,\sigma\in\mathscr{B}(\Delta,\gamma)} \left(\frac{\mathrm{e}^{rc'x(\sigma)}}{\mathrm{e}^{rc'x(\sigma^*)}}\right)^t \frac{U_\sigma(tb,rc)}{\mu(\sigma)}\right]^{1/r}$$

and from (5.17)–(5.18), setting $\delta_\sigma := c'x^* - c'x(\sigma) > 0$ and $u := \mathrm{e}^{r/q}$,

$$\widehat{f}_d(ty,rc)^{1/r} = \mathrm{e}^{tc'x^*}\left[\frac{U_{\sigma^*}(ty,rc)}{\mu(\sigma^*)} + \sum_{x(\sigma)\neq x^*:\,\sigma\in\mathscr{B}(\Delta,\gamma)} u^{-tq\delta_\sigma}\frac{P_{\sigma ty}(u)}{Q_{\sigma ty}(u)}\right]^{1/r}.$$

Observe that $c'x(\sigma^*) - c'x(\sigma) > 0$ whenever $\sigma \neq \sigma^*$ because $\Omega(y)$ is simple if $y \in \gamma$ and c is regular. Indeed, as x^* is an optimal vertex of the LP problem $\mathbf{P}$, the reduced costs $(c_k - \pi_{\sigma^*}A_k)$ $(k \notin \sigma^*)$ with respect to the optimal basis σ^* are all nonpositive and, in fact, strictly negative because c is regular (see Section 2.3). Therefore, the term

$$\sum_{\text{vertex}\,x(\sigma)\neq x^*} u^{-tq\delta_\sigma}\frac{P_{\sigma ty}(u)}{Q_{\sigma ty}(u)}$$

is negligible for t sufficiently large, when compared with $U_{\sigma^*}(tu,rc)$. This is because the degrees of $P_{\sigma ty}$ and $Q_{\sigma ty}$ depend on ty but *not* on the magnitude of ty (see (5.17)–(5.18)), and they are uniformly bounded in ty. Hence, taking limit as $r\to\infty$ yields

$$\mathrm{e}^{f_d(ty,c)} = \lim_{r\to\infty}\left[\frac{\mathrm{e}^{rtc'x(\sigma^*)}}{\mu(\sigma^*)}U_{\sigma^*}(ty,rc)\right]^{1/r} = \mathrm{e}^{tc'x(\sigma^*)}\lim_{r\to\infty}U_{\sigma^*}(ty,rc)^{1/r},$$

from which (5.5) follows easily.

Finally, the periodicity is coming from the term $\mathrm{e}^{2i\pi ty}(g)$ in (4.15) (with ty in lieu of y) for $g \in G(\sigma^*)$. The period is then the order of $G(\sigma^*)$, that is, the volume of the convex polytope $\{\sum_{j\in\sigma^*} t_jA_j \,|\, 0 \le t_j \le 1\}$, normalized so that volume $(\mathbb{R}^m/\Lambda) = 1$, i.e., $\mu(\sigma)$. □

Proof of Proposition 5.1

With $U_{\sigma^*}(y,c)$ as in (4.15), one has $\pi_{\sigma^*} = (\lambda^*)'$ and so

$$\mathrm{e}^{-\pi_{\sigma^*}A_k}\mathrm{e}^{-2i\pi A_k}(g) = \mathrm{e}^{-A_k'\lambda^*}\mathrm{e}^{-2i\pi A_k'\theta_g} = z_g^{-A_k}, \quad g \in G(\sigma^*).$$

Next, using $c'x(\sigma^*) = y'\lambda^*$,

$$\mathrm{e}^{c'x(\sigma^*)}\mathrm{e}^{2i\pi y}(g) = \mathrm{e}^{y'\lambda^*}\mathrm{e}^{2i\pi y'\theta_g} = z_g^y, \quad g \in G(\sigma^*).$$

Therefore,

$$\frac{1}{\mu(\sigma^*)}\mathrm{e}^{c'x(\sigma)}U_{\sigma^*}(y,c) = \frac{1}{\mu(\sigma^*)}\sum_{g\in G(\sigma^*)}\frac{z_g^y}{(1-z_g^{-A_k}\mathrm{e}^{c_k})} = R_{\sigma^*}(z_g,1),$$

and (5.14) follows from (5.4) because, with rc in lieu of c, z_g becomes $z_{gr} = \mathrm{e}^{r\lambda^*}\mathrm{e}^{2i\pi\theta_g}$ (only the modulus changes).

Next, as only the modulus of z_g is involved in (5.15), we have $|z_{gr}| = \mathrm{e}^{r\lambda^*}$ for all $g \in G(\sigma^*)$, so that

$$\frac{1}{\mu(\sigma^*)}\sum_{g\in G(\sigma^*)}\frac{|z_{gr}|^y}{\prod_{k\notin\sigma^*}(1-|z_{gr}|^{-A_k}\mathrm{e}^{rc_k})} = \frac{\mathrm{e}^{ry'\lambda^*}}{\prod_{k\notin\sigma^*}(1-\mathrm{e}^{r(c_k-A_k'\lambda^*)})},$$

and, as $r\to\infty$,

$$\frac{\mathrm{e}^{ry'\lambda^*}}{\prod_{k\notin\sigma^*}(1-\mathrm{e}^{r(c_k-A_k'\lambda^*)})} \approx \mathrm{e}^{ry'\lambda^*},$$

because $(c_k - A_k'\lambda^*) < 0$ for all $k \notin \sigma^*$. Therefore,

$$\lim_{r\to\infty}\frac{1}{r}\ln\left[\frac{\mathrm{e}^{ry'\lambda^*}}{\prod_{k\notin\sigma^*}(1-\mathrm{e}^{r(c_k-A_k'\lambda^*)})}\right] = y'\lambda^* = f(y,c),$$

the desired result. □

5.5 Notes

Most of the material of this chapter is taken from Lasserre [87, 88]. We have seen that several duality results can be derived from the generating function $\widehat{F}_d$ of the counting problem $\mathbf{I}_d$ associated with the integer program $\mathbf{P}_d$. However, the asymptotic results of Corollary 5.1 had been already obtained in Gomory [58] via a (primal) algebraic approach. In the next chapter we analyze Brion and Vergne's discrete formula and make the link with the Gomory relaxations introduced in the algebraic approach of Gomory [58].

Part III
Duality

Chapter 6
Duality and Gomory Relaxations

The last part of the book is mainly concerned with duality results for the integer programming poblem $\mathbf{P}_d$. We relate the old agebraic concept of Gomory relaxation with results of previous chapters. We also provide some new duality results and relate them to superadditivity.

6.1 Introduction

We pursue our investigation on the integer program $\mathbf{P}_d$ in (5.1) from a duality point of view, initiated in previous chapters. The main goal of this chapter is to provide additional insights into $\mathbf{P}_d$ and relate some results already obtained in Chapter 5 with the so-called *Gomory relaxations*, an algebraic approach introduced by Gomory [58] in the late 1960s further studied in Wolsey [133, 134, 136], and more recently by Gomory et al. [60], Thomas [128], and Sturmfels and Thomas [126] (see also Aardal et al. [1]).

In particular, we will show that Brion and Vergne's formula (2.15) is strongly related to Gomory relaxations. Interestingly, the Gomory relaxations also have a nice interpretation in terms of *initial ideals* in the algebraic approaches developed in Sturmfels and Thomas [126] and Thomas [128].

Next, as already mentioned, most duality results available for integer programs have been obtained via the use of *superadditive* functions as in, e.g., Wolsey [136], and the smaller class of *Chvátal* and *Gomory* functions as in, e.g., Blair and Jeroslow [23] (see also Schrijver [121, pp. 346–353] and the many references therein). However, Chvátal and Gomory functions are only defined implicitly from their properties, and the resulting dual problems defined in, e.g., [23, 71, 136] are essentially conceptual in nature; rather, Gomory functions are used to generate cutting planes for the (primal) integer program $\mathbf{P}_d$. For instance, a dual problem of $\mathbf{P}_d$ is the optimization problem

$$\min_{f\in\Gamma}\{f(y)|f(A_j)\geq c_j, \quad j=1,\ldots,n\}, \tag{6.1}$$

where Γ is the set of all *superadditive* functions $f:\mathbb{R}^m$ (or $\mathbb{Z}^m$)$\to\mathbb{R}$, with $f(0)=0$ (called *price functions* in Wolsey [134]). In principle, this dual problem permits us to retrieve many ingredients of linear programming (e.g., weak and strong duality and complementarity slackness) and to derive qualitative postoptimality analysis (see, e.g., Wolsey [134]). However, as already said, the dual

J.B. Lasserre, *Linear and Integer Programming vs Linear Integration and Counting*,
Springer Series in Operations Research and Financial Engineering, DOI 10.1007/978-0-387-09414-4_6,

problem (6.1) is rather conceptual as the constraint set in (6.1) is not easy to handle, even if it may reduce to a (huge) linear program. For instance, the characterization of the integer hull of the convex polyhedron $\{x \in \mathbb{R}^n | Ax \leq y, x \geq 0\}$ via the functions $f \in \Gamma$ in Wolsey [134, Theor. 6] is mainly of theoretical nature.

We already mentioned in Chapter 4 that the approach we follow is *dual* to the algebraic methods just mentioned. Indeed, recall that the variables of interest associated with the constraints $Ax = y$ are the analogues of the usual *dual* variables in linear programming, except they are now in $\mathbb{C}^m$ rather than in $\mathbb{R}^m$. So, if in the primal algebraic approaches described in Sturmfels and Thomas [126] and Thomas [128], integer programming appears as an *arithmetic refinement* of linear programming, in the present dual approach, integer programming appears as a complexification (in $\mathbb{C}^m$) of the associated LP dual (in $\mathbb{R}^m$). That is, restricting the primal LP (in $\mathbb{R}^n$) to the integers $\mathbb{N}^n$ induces enlarging the dual LP (in $\mathbb{R}^m$) to $\mathbb{C}^m$.

In this Chapter latter statement is clarified and a dual problem $\mathbf{P}_d^*$ is made explicit. It can be interpreted as an analogue of the linear program dual $\mathbf{P}^*$ of $\mathbf{P}$ in the sense that it is obtained in a similar fashion, by using an analogue of the Legendre–Fenchel transform in which the dual variables z are now in $\mathbb{C}^m$. If z is replaced with its vector $|z|$ of component moduli, we retrieve the usual Legendre–Fenchel transform, and thus, the usual dual $\mathbf{P}^*$.

6.2 Gomory relaxations

With $A \in \mathbb{Z}^{m\times n}$, $y \in \mathbb{Z}^m$, $c \in \mathbb{Q}^n$, consider the integer program $\mathbf{P}_d$ in (5.1) and its associated linear program $\mathbf{P}$, with respective optimal values $f_d(y,c)$ and $f(y,c)$.

Let σ be an optimal basis of $\mathbf{P}$, and let A be partitioned into $[A_\sigma | A_N]$ with $A_\sigma \in \mathbb{Z}^{m\times m}$ being the matrix basis, and $A_N \in \mathbb{Z}^{m\times(n-m)}$. With $c_\sigma := \{c_j\}_{j\in\sigma}$, let $\pi_\sigma := c_\sigma A_\sigma^{-1} \in \mathbb{R}^m$ be the dual vector associated with σ, and let $x(\sigma) = [x_\sigma, x_N] \in \mathbb{R}^n$ be an optimal vertex. Consider the integer program:

$$\mathrm{IP}_\sigma(y): \quad \begin{cases} \max\limits_{x,v} & \sum_{k\notin\sigma} (c_k - \pi_\sigma A_k) x_k \\ \text{s.t.} & A_\sigma v + \sum_{k\notin\sigma} A_k x_k = y, \\ & v \in \mathbb{Z}^m,\ x_k \in \mathbb{N} \quad \forall k \notin \sigma, \end{cases} \tag{6.2}$$

with optimal value denoted by $\max \mathrm{IP}_\sigma(y)$. Notice that $\max \mathrm{IP}_\sigma(y) \leq 0$ because σ being an optimal basis makes the *reduced costs* $\{c_k - \pi_\sigma A_k\}_{k\notin\sigma}$ all nonpositive. It is called a *relaxation* of $\mathbf{P}_d$ because $\mathrm{IP}_\sigma(y)$ is obtained from $\mathbf{P}_d$ by relaxing the nonnegativity constraints on the variables x_j with $j \in \sigma$. Therefore,

$$f_d(y,c) \leq c'x(\sigma) + \max \mathrm{IP}_\sigma(y) \leq c'x(\sigma), \qquad \forall y \in \mathbb{Z}^m.$$

Next, the integer program $\mathrm{IP}_\sigma(y)$ can be written in the equivalent form:

$$\begin{cases} \max_x & \sum_{k\notin\sigma} (c_k - \pi_\sigma A_k)\, x_k \\ \text{s.t.} & \sum_{k\notin\sigma} (A_\sigma^{-1} A_k)\, x_k = A_\sigma^{-1} y \,(\mathrm{mod}\, 1), \\ & x_k \in \mathbb{N} \quad \forall k \notin \sigma \end{cases} \tag{6.3}$$

a particular instance of the more general *group problem*

$$\mathrm{IP}_G(g_0): \quad \max_{x\in\mathbb{N}^s} \left\{ c'x \,|\, \sum_{k=1}^{s} g_k x_k = g_0 \text{ in } G \right\}, \tag{6.4}$$

where $(G,+)$ is a finite abelian group, $\{g_k\} \subset G$, and $g_k x_k$ stands for $g_k + \cdots + g_k$, x_k times. Indeed, in view of (6.3) take $G := \Lambda/\mathbb{Z}^m$ with Λ being the group $A_\sigma^{-1}(\mathbb{Z}^m)$, so that $|G| = |A_\sigma|$. That is why (6.2) (or (6.3)) is called the *Gomory (asymptotic) group relaxation.* In fact (with the notation $a|b$ standing for a divides b, when $a,b \in \mathbb{Z}$) we have:

Proposition 6.1. *Every finite abelian group is isomorphic to the group $\mathbb{Z}_{\delta_1} \times \cdots \times \mathbb{Z}_{\delta_p}$ of integer p-vectors with addition modulo $(\delta_1,\dots,\delta_p)$, with $\delta_i \in \mathbb{Z}_+ \setminus \{0,1\}$ and $\delta_1|\delta_2|\cdots|\delta_n$.*

See for instance Aardal et al. [1]. Solving the group problem $\mathrm{IP}_G(g_0)$ reduces to solving a longest path problem in a graph $\tilde{G}$ with $|G|$ nodes. From each node $g \in \tilde{G}$ there is an arc $g \to g + g_j$ with associated reward $c_j - \pi_\sigma A_j \geq 0$, and one wishes to find the longest path from 0 to $g_0 \in \tilde{G}$. This can be achieved in $O((n-m)|G|)$ operations. In addition, there is an optimal solution $x \in \mathbb{N}^s$ of (6.4) such that $\prod_{k=1}^{s}(1+x_k) \leq |G|$; see [1, p. 61].

Of course if an optimal solution $x_N^* = \{x_j^*\}_{j\notin\sigma}$ of the group problem (6.3) satisfies $x_\sigma^* := A_\sigma^{-1}y - A_\sigma^{-1}A_N x_N^* \geq 0$ then $[x_\sigma^*, x_N^*] \in \mathbb{N}^n$ is an optimal solution of $\mathbf{P}_d$, and the Gomory relaxation (6.2) is said to be *exact.* This eventually happens when y is large enough (whence the name Gomory "asymptotic" group relaxation).

The *corner polyhedron* $\mathrm{CP}(\sigma)$ associated with $\mathbf{P}_d$ is the convex hull

$$\mathrm{CP}(\sigma) := \mathrm{conv}\,\{\, x_N \in \mathbb{Z}_+^{n-m} \,|\, A_\sigma^{-1} y - A_\sigma^{-1} A_N x_N \in \mathbb{Z}^m \,\} \tag{6.5}$$

of the feasible set of (6.3). In particular, it contains the convex hull

$$\mathrm{conv}\,\{\, x_N \in \mathbb{Z}_+^{n-m} \,|\, A_\sigma^{-1} y - A_\sigma^{-1} A_N x_N \in \mathbb{Z}_+^m \,\}$$

of the projection on the nonbasic variables of the feasible set of $\mathbf{P}_d$. An important property of $\mathrm{CP}(\sigma)$ is that its facets are *cutting planes* of the integer programming problem $\mathbf{P}_d$.

More generally, Gomory considered the feasible set of the group problem $\mathrm{IP}_G(g_0)$ when *all* elements $g \in G$ appear, i.e., the group constraint in (6.4) reads $\sum_{g\in G} g x_g = g_0$. Its convex hull $\mathrm{P}(G, g_0)$ is called the *master polyhedron* associated with G and $g_0 \neq 0$, and its facet-defining inequalities can be characterized in terms of subadditive functions; namely, as described in the following:

Proposition 6.2. *Let $0 \neq g_0 \in G$, and let $\pi \in \mathbb{R}_+^{|G|}$ with $\pi_0 = 0$. The inequality $\sum_{g\in G} \pi_g x_g \geq 1$ is facet-defining for $\mathrm{P}(G, g_0)$ if and only if*

$$\begin{cases} \pi_{g_1} + \pi_{g_2} \geq \pi_{g_1+g_2} & \forall g_1, g_2 \in G \setminus \{0, g_0\}, \\ \pi_g + \pi_{g_0-g} = \pi_{g_0} = 1 & \forall g \in G \setminus \{0\}. \end{cases} \tag{6.6}$$

See, e.g., Gomory et al. [60] or Aardal et al. [1]. In fact, the facets of $\mathrm{P}(G, g_0)$ are also cutting planes of corner polyhedra associated with $\mathrm{IP}(G, g_0)$ in (6.4) when only a subset of G appears in the group constraint.

Among the many facets, some are likely to be "larger" (hence more important) than others, and so one would like to detect them without knowing all the facets. This is done by Gomory's *shooting theorem*, which asserts that the facet of $\mathrm{P}(G, g_0)$ hit by a random direction v is obtained by solving the linear program that minimizes $\pi' v$ over the feasible set (6.6). In addition, knowledge of facets for corner polyhedra with small group G also permits us to derive cutting planes for integer programs of any size. This approach is developed in Gomory et al. [60].

6.3 Brion and Vergne's formula and Gomory relaxations

Consider the counting problem $\mathbf{I}_d$ (5.2) with value $\widehat{f_d}(y, c)$, associated with the integer program $\mathbf{P}_d$.

With the same notation and definitions as in Section 2.3, let $\Lambda = A(\mathbb{Z}^n)$ and let $c \in \mathbb{R}^n$ be regular with $-c$ in the interior of the dual cone $(\mathbb{R}^n_+ \cap V)^*$. For a basis σ, let $\mu(\sigma)$ be the volume of the convex polytope $\{\sum_{j\in\sigma} A_j t_j, 0 \leq t_j \leq 1, \forall j \in \sigma\}$, normalized so that $\mathrm{vol}(\mathbb{R}^n/\Lambda) = 1$. Recall that for a chamber γ, and every $y \in \Lambda \cap \overline{\gamma}$, Brion and Vergne's discrete formula (4.14) reads

$$\widehat{f_d}(y, c) = \sum_{x(\sigma):\, \sigma \in \mathscr{B}(\Delta, \gamma)} \frac{\mathrm{e}^{c' x(\sigma)}}{\mu(\sigma)} U_\sigma(y, c)$$

$$\text{with} \quad U_\sigma(y, c) = \sum_{g \in G(\sigma)} \frac{\mathrm{e}^{2i\pi y}(g)}{\prod_{k \notin \sigma} (1 - \mathrm{e}^{-2i\pi A_k}(g) \mathrm{e}^{(c_k - \pi_\sigma A_k)})}. \tag{6.7}$$

In (6.7), $G(\sigma)$ is the finite abelian group $(\oplus_{j\in\sigma} \mathbb{Z} A_j)^* / \Lambda^*$ of order $\mu(\sigma)$ (where Λ^* is the dual lattice of Λ); see Section 4.3. We next make explicit the term $U_\sigma(y, c)$ and relate it with Gomory relaxations.

The optimal value of $\mathbf{P}_d$

We first refine the characterization (6.7). We assume for convenience that $c \in \mathbb{Q}^n$, but the result still holds for $c \in \mathbb{R}^n$ (see Remark 6.1). Let $y \in \Lambda$. Given a vertex $x(\sigma)$ of $\Omega(y)$, let $\pi_\sigma A_\sigma = c_\sigma$ (with $c_\sigma = \{c_j\}_{j\in\sigma}$), and define

$$S_\sigma := \{k \notin \sigma \mid A_k \notin \oplus_{j\in\sigma} A_j \mathbb{Z}\} \tag{6.8}$$

and

$$
\begin{aligned}
M_\sigma^+ &:= \{k \notin \sigma | (c_k - \pi_\sigma A_k) > 0\}, \\
M_\sigma^- &:= \{k \notin \sigma | (c_k - \pi_\sigma A_k) < 0\}.
\end{aligned}
\tag{6.9}
$$

When c is regular, then M_σ^+, M_σ^- define a partition of $\{1,\ldots,n\} \setminus \sigma$. Note also that for the (unique) optimal vertex $x(\sigma^*)$ of the linear program $\mathbf{P}$, $M_{\sigma^*}^+ = \emptyset$. Finally, for every $k \in S_\sigma$, denote by $s_{k\sigma} \in \mathbb{N}$ the smallest integer such that $s_{k\sigma} A_k \in \oplus_{j\in\sigma} A_j \mathbb{Z}$.

Lemma 6.1. *Let $c \in \mathbb{Q}^n$ be regular with $-c \in (\mathbb{R}_+^n \cap V)^*$, and $y \in \Lambda$. Let $q \in \mathbb{N}$ be large enough to ensure that $qc \in \mathbb{Z}^m$ and $q(c_k - \pi_\sigma A_k) \in \mathbb{Z}$ for all $\sigma \in \mathscr{B}(\Delta), k \notin \sigma$, and let $u := \mathrm{e}^{r/q}$, $r \in \mathbb{R}$. Let $U_\sigma(y,c)$ be as in (6.7). Then*

(i) $U_\sigma(y,rc)$ can be written as

$$
U_\sigma(y,rc) = \frac{P_{\sigma y}}{Q_{\sigma y}} \tag{6.10}
$$

for two Laurent polynomials $P_{\sigma y}, Q_{\sigma y} \in \mathbb{R}[u,u^{-1}]$. In addition, the maximal algebraic degree of $P_{\sigma y}$ is given by

$$
\begin{cases}
\quad q \max \sum_{k\in S_\sigma} (c_k - \pi_\sigma A_k) x_k \\
\text{s.t.} \quad A_\sigma v + \sum_{k\in S_\sigma} A_k x_k = y, \\
\quad v \in \mathbb{Z}^m,\ x_k \in \mathbb{N},\ x_k < s_{k\sigma} \quad \forall k \in S_\sigma,
\end{cases}
\tag{6.11}
$$

whereas the maximal algebraic degree of $Q_{\sigma y}$ is given by

$$
q \sum_{\substack{k\notin S_\sigma \\ k\in M_\sigma^+}} (c_k - \pi_\sigma A_k) + q s_{k\sigma} \sum_{\substack{k\in S_\sigma \\ k\in M_\sigma^+}} (c_k - \pi_\sigma A_k). \tag{6.12}
$$

(ii) As a function of the variable $u := \mathrm{e}^{r/q}$, and when $r \to \infty$,

$$
\mathrm{e}^{rc'x(\sigma)} U_\sigma(y,rc) \approx u^{qc'x(\sigma) + \deg P_{\sigma y} - \deg Q_{\sigma y}}, \tag{6.13}
$$

where "deg " denotes the algebraic degree (i.e., the largest power, sign included).

A detailed proof is postponed until Section 6.6.

Remark 6.1. Lemma 6.1 is also valid if $c \in \mathbb{R}^n$ instead of $\mathbb{Q}^n$. But this time, $P_{\sigma y}$ and $Q_{\sigma y}$ in (6.10) are no longer Laurent polynomials.

As a consequence of Lemma 6.1, we get

Corollary 6.1. *Let c be regular with $-c \in (\mathbb{R}^n_+ \cap V)^*$, $y \in \Lambda \cap \overline{\gamma}$, and let $x(\sigma)$, $\sigma \in \mathscr{B}(\Delta, \gamma)$, be a vertex of $\Omega(y)$. Then with $U_\sigma(y,c)$ as in (6.7),*

$$c'x(\sigma) + \lim_{r \to \infty} \frac{1}{r} \ln U_\sigma(y, rc) = c'x(\sigma) + \max \mathrm{IP}_\sigma(y), \tag{6.14}$$

where $\max \mathrm{IP}_\sigma(y)$ *is the optimal value of the integer program* $\mathrm{IP}_\sigma(y)$*:*

$$\left\{ \begin{array}{ll} \displaystyle\max_{x,v} & \displaystyle\sum_{k \in M_\sigma^+} (c_k - \pi_\sigma A_k)(x_k - s_{k\sigma}) + \sum_{k \in M_\sigma^-} (c_k - \pi_\sigma A_k) x_k \\ \text{s.t.} & A_\sigma v + \sum_{k \notin \sigma} A_k x_k = y, \\ & v \in \mathbb{Z}^m, x_k \in \mathbb{N}, x_k < s_{k\sigma} \quad \forall k \notin \sigma. \end{array} \right. \tag{6.15}$$

For a proof see Section 6.6.

The link with Gomory relaxations

When σ is the optimal basis σ^* of the linear program **P**, the integer program $\mathrm{IP}_\sigma(y)$ in Corollary 6.1 reads

$$\mathrm{IP}_{\sigma^*}(y) \left\{ \begin{array}{ll} \displaystyle\max_{x,v} & \sum_{k \notin \sigma^*} (c_k - \pi_{\sigma^*} A_k) x_k \\ \text{s.t.} & A_{\sigma^*} v + \sum_{k \notin \sigma^*} A_k x_k = y, \\ & v \in \mathbb{Z}^m, x_k \in \mathbb{N}, x_k < s_{k\sigma^*} \quad \forall k \notin \sigma^* \end{array} \right.$$

(because $M_\sigma^+ = \emptyset$). Observe that all the variables x_k, $k \notin S_{\sigma^*}$, must be zero because of $x_k < s_{k\sigma^*} = 1$.

Next, the constraint $x_k < s_{k\sigma^*}$ in $\mathrm{IP}_{\sigma^*}(y)$ can be removed. Indeed, if in a solution (v,x) of $\mathrm{IP}_{\sigma^*}(y)$ some variable x_k can be written $ps_{k\sigma^*} + r_k$ for some $p, r_k \in \mathbb{N}$ with $p > 0$, then one may replace x_k with $\tilde{x}_k := r_k$ and obtain a better value because $(c_k - \pi_{\sigma^*} A_k) < 0$ and $pA_k s_{k\sigma^*} = A_{\sigma^*} w$ for some $w \in \mathbb{Z}^m$. Therefore, $\mathrm{IP}_{\sigma^*}(y)$ is nothing else than the Gomory asymptotic group relaxation defined in (6.3) with $\sigma = \sigma^*$.

Observe that for every basis $\sigma \neq \sigma^*$ of the linear program **P**, $\max \mathrm{IP}_\sigma(y) < 0$ and so,

$$c'x(\sigma) + \max \mathrm{IP}_\sigma(y) < c'x(\sigma) \qquad \sigma \neq \sigma^*. \tag{6.16}$$

In addition, with $\rho_\sigma := \sum_{k \in M_\sigma^+} (c_k - \pi_\sigma A_k) s_{k\sigma} > 0$, one may rewrite (6.15) as

$$\begin{cases} -\rho_\sigma + \max_{x,v} \sum_{k\notin\sigma}(c_k - \pi_\sigma A_k)x_k \\ \text{s.t. } \sum_{k\notin\sigma}(A_\sigma^{-1}A_k)x_k = A_\sigma^{-1}y(\text{mod } 1), \\ \quad x_k \in \{0,1,\ldots,s_{k\sigma}-1\} \quad \forall k \notin \sigma, \end{cases} \tag{6.17}$$

which is a *group relaxation* problem as defined in (6.3) with a bound constraint $x_k < s_{k\sigma}$ on the variables. Indeed, when the cost coefficient of x_k is positive, one needs this bound; otherwise the relaxation has infinite optimal value because $A_\sigma^{-1}A_k s_{k\sigma} \equiv 0$ in $G = A_\sigma^{-1}(\mathbb{Z}^m)/\mathbb{Z}^m$.

Corollary 6.1 shows how this group relaxation concept naturally arises from a dual point of view that considers the counting function $\widehat{f_d}(y,c)$. Here, and as in Thomas [128], it is defined for an arbitrary basis σ, and not necessarily for the optimal basis σ^* of the linear program **P**, as in Wolsey [134].

The group relaxations $\mathrm{IP}_\sigma(y)$ in (6.15), or equivalently in (6.17), are defined for all *feasible bases* σ of the LP problem **P** associated with $\mathbf{P}_d$, whereas the (extended) group relaxations in Hosten and Thomas [67] and Thomas [128] are defined with respect to the feasible bases σ of the LP *dual* $\mathbf{P}^*$ of **P**. So our former *primal group relaxations* are bounded because of the constraint $x_k < s_{k\sigma}$ for all $k \in M_\sigma^+$, whereas the latter *dual group relaxations* of Hosten and Thomas are bounded because $(c_k - \pi_\sigma A_k) < 0$ for all $k \notin \sigma$, i.e., $M_\sigma^+ = \emptyset$; therefore, and because $M_\sigma^+ = \emptyset$, the latter do not include the bound constraints $x_k < s_{k\sigma}$, and the cost function does not include the term $-(c_k - \pi_\sigma A_k)s_{k\sigma}$. In addition, in the *extended* dual group relaxations of Hosten and Thomas [67] associated with a basis σ, one enforces the nonnegativity of x_k for some indices $k \in \sigma$ (as in the extended group relaxations of Wolsey [134] for the optimal basis σ^*). Finally, note that the bound constraint $x_k < s_{k\sigma}$ in (6.15) is not added artificially; it comes from a detailed analysis of the leading term of the rational fraction $P_{\sigma y}(u)/Q_{\sigma y}(u)$ in (6.10) as $u\to\infty$. In particular, the constant term $\sum_{k\in M_\sigma^+}(c_k - \pi_\sigma A_k)s_{k\sigma}$ is the degree of the leading term of $Q_{\sigma y}(u)$; see (6.12) ($s_{k\sigma} = 1$ if $k \notin S_\sigma$). This is why, when we look at the leading power in (6.13), this term appears with a minus sign in (6.15).

One may also define what could be called *extended* primal group relaxations, that is, primal group relaxations $\mathrm{IP}_\sigma(y)$ in (6.15), with *additional* nonnegativity constraints on some components of the vector v. They would be the primal analogues of the extended dual group relaxations of Hosten and Thomas. Roughly speaking, enforcing nonnegativity conditions on some components of the vector v in (6.15) amounts to looking at *nonleading* terms of $P_{\sigma y}(u)/Q_{\sigma y}(u)$ (see Remark 6.2 below).

Remark 6.2. Let us go back to the definition (6.7) of $U_\sigma(y,c)$, that is (with $u = e^{1/q}$), the compact formula

$$U_\sigma(y,c) = \sum_{g\in G(\sigma)} \frac{\mathrm{e}^{2i\pi y}(g)}{\prod_{k\notin\sigma}(1 - \mathrm{e}^{-2i\pi A_k}(g)u^{q(c_k - \pi_\sigma A_k)})}. \tag{6.18}$$

Developing and writing $U_\sigma(y,c) = P_{\sigma b}(u)/Q_{\sigma b}(u)$, with $P_{\sigma y}, Q_{\sigma y} \in \mathbb{R}[u,u^{-1}]$, the Laurent polynomial $P_{\sigma y}$ *encodes* all the values v of the feasible solutions of the group relaxation $\mathrm{IP}_\sigma(y)$ in the powers of its monomials u^v and the number of solutions with value v, in the coefficient of u^v (see Section 6.6). So (6.18) is a compact encoding of the group relaxation $\mathrm{IP}_\sigma(y)$.

We now obtain

Theorem 6.1. *Let c be regular with $-c \in (\mathbb{R}^n_+ \cap V)^*$, and $y \in \Lambda \cap \overline{\gamma}$. Assume that $Ax = y$ has a solution $x \in \mathbb{N}^n$. If the max in (5.3) is attained at a unique vertex $x(\sigma^*)$ of $\Omega(y)$, then $\sigma^* \in \mathscr{B}(\Delta, \gamma)$ is an optimal basis of the linear program* **P**, *and*

$$\begin{aligned} f_d(y,c) &= c'x(\sigma^*) + \max \mathrm{IP}_{\sigma^*}(y) \\ &= c'x(\sigma^*) + \left[\begin{array}{l} \max \sum_{k \notin \sigma^*} (c_k - \pi_{\sigma^*} A_k) x_k \\ A_\sigma v + \sum_{k \notin \sigma^*} A_k x_k = y \\ v \in \mathbb{Z}^m,\, x_k \in \mathbb{N} \quad \forall k \notin \sigma^*. \end{array} \right. \end{aligned} \tag{6.19}$$

Equivalently, the gap between the discrete and continuous optimal values is given by

$$f_d(y,c) - f(y,c) = \max \mathrm{IP}_{\sigma^*}(y). \tag{6.20}$$

Proof. Let $\sigma^* \in \mathscr{B}(\Delta, \gamma)$ be an optimal basis of the linear program **P**, with corresponding optimal solution $x(\sigma^*) \in \mathbb{R}^n_+$. Let x^* be an optimal solution of $\mathbf{P}_d$ and let $v_{\sigma^*} := \{x^*_j\}_{j \in \sigma^*} \in \mathbb{N}^m$. The vector $(v_{\sigma^*}, \{x^*_k\}_{k \notin \sigma^*}) \in \mathbb{N}^n$ is a feasible solution to $\mathrm{IP}_{\sigma^*}(y)$. Moreover,

$$\sum_{j=1}^n c_j x^*_j = c'x(\sigma^*) + \sum_{j \notin \sigma^*} (c_k - \pi_{\sigma^*} A_k) x^*_k. \tag{6.21}$$

Therefore, if the max in (5.3) is attained at a unique vertex, then by (5.4),

$$\begin{aligned} f_d(y,c) &= \max_{x(\sigma):\ \text{vertex of } \Omega(y)} \left[c'x(\sigma) + \lim_{r \to \infty} \frac{1}{r} \ln U_\sigma(y, rc) \right] \\ &\geq c'x(\sigma^*) + \lim_{r \to \infty} \frac{1}{r} \ln U_{\sigma^*}(y, rc) \\ &= c'x(\sigma^*) + \max \mathrm{IP}_{\sigma^*}(y) \\ &\geq c'x(\sigma^*) + \sum_{k \notin \sigma^*} (c_k - \pi_{\sigma^*} A_k) x^*_k \\ &= \sum_{j=1}^n c_j x^*_j = f_d(y,c), \end{aligned}$$

and so the max in (5.3) is necessarily attained at $\sigma = \sigma^*$. □

As noted in Gomory [58], when the group relaxation $\mathrm{IP}_{\sigma^*}(y)$ provides an optimal solution $x^* \in \mathbb{N}^n$ of $\mathbf{P}_d$, then x^* is obtained from an optimal solution $x(\sigma^*)$ of the linear program **P** and a *periodic* correction term $\sum_{k \notin \sigma^*} (c_k - \pi_{\sigma^*} A_k) x^*_k$. Indeed, for all $\tilde{y} := y + A_{\sigma^*} v$, $v \in \mathbb{Z}^m$, the group relaxation $\mathrm{IP}_{\sigma^*}(\tilde{y})$ has same optimal value as $\mathrm{IP}_{\sigma^*}(y)$.

We also obtain the following sufficient condition on the data of $\mathbf{P}_d$ to ensure that the group relaxation $\mathrm{IP}_{\sigma^*}(y)$ provides an optimal solution of $\mathbf{P}_d$.

Corollary 6.2. *Let c be regular with $-c \in (\mathbb{R}^n_+ \cap V)^*$, and $y \in \Lambda \cap \overline{\gamma}$. Let $x(\sigma^*)$ be an optimal vertex of the linear program* **P** *with optimal basis $\sigma^* \in \mathscr{B}(\Delta,\gamma)$. If*

$$c'x(\sigma^*) + \sum_{k \notin \sigma^*} (c_k - \pi_{\sigma^*} A_k)(s_{k\sigma^*} - 1) > c'x(\sigma) - \sum_{\substack{k \notin \sigma \\ k \in M_\sigma^+}} (c_k - \pi_\sigma A_k) \tag{6.22}$$

for every vertex $x(\sigma)$ of $\Omega(y)$, $\sigma \in \mathscr{B}(\Delta,\gamma)$, then the "max" in (5.3) is attained at $\sigma = \sigma^*$ and

$$f_d(y,c) = c'x(\sigma^*) + \max \mathrm{IP}_{\sigma^*}(y). \tag{6.23}$$

Proof. The result follows because the left-hand side of (6.22) is a lower bound on $\mathrm{IP}_{\sigma^*}(y)$, whereas the right-hand side is an upper bound on the optimal value of $\mathrm{IP}_\sigma(y)$. □

Note that for $t \in \mathbb{N}$ sufficiently large and $y := ty$, the condition (6.22) is certainly true. So for sufficiently large $t \in \mathbb{N}$, the optimal value of $\mathbf{P}_d$ (where $Ax = ty$) is given by (6.23).

The nondegeneracy property

When the group relaxation $\mathrm{IP}_{\sigma^*}(y)$ does not provide an optimal solution of $\mathbf{P}_d$, Theorem 6.1 states that *necessarily* the max in (5.3) is attained at *several* bases σ. It is not just related to the fact that at some optimal solution (v,x) of $\mathrm{IP}_{\sigma^*}(y)$, the vector $v \in \mathbb{Z}^m$ has some negative components. There is at least *another* group relaxation with basis $\sigma \neq \sigma^*$, and such that $c'x(\sigma) + \max \mathrm{IP}_\sigma(y)$ is also the maximum in (5.3).

On the other hand, when the uniqueness property of the max in (5.3) holds, then one may qualify $\sigma^* \in \mathscr{B}(\Delta,\gamma)$ as the unique *optimal basis of the integer program* $\mathbf{P}_d$, in the sense that the group relaxation $\mathrm{IP}_{\sigma^*}(y)$ is the only one to provide this max (and hence, an optimal solution of $\mathbf{P}_d$). Equivalently, the uniqueness of the max in (5.3) is also the uniqueness of the max in

$$\max_{x(\sigma):\sigma \in \mathscr{B}(\Delta,\gamma)} \left\{ c'x(\sigma) + \max \mathrm{IP}_\sigma(y) \right\}.$$

This uniqueness property of the max in (5.3) is the discrete analogue of the linear programming *nondegeneracy* property. Indeed, an optimal basis is not unique if the optimal vertex of the dual is *degenerate* (that is, there are two different optimal bases of the dual LP with same vertex; observe that when $y \in \gamma$ then $\Omega(y)$ is a *simple* polyhedron (see Brion and Vergne [27, Prop. p. 818]) and the nondegeneracy property holds for the primal.

Thus, when $y \in \gamma$ (so that the optimal vertex $x^* \in \mathbb{R}^n_+$ is nondegenerate), and $c \in \mathbb{R}^n$ is regular, then σ^* is the unique optimal basis of the linear program **P**. As $A_k \in \oplus_{j \in \sigma} A_j \mathbb{R}$ for all $k \notin \sigma$ we may set $s_{k\sigma} := 1$ for all $k \notin \sigma$, and all σ so that the linear program

$$\mathrm{LP}_{\sigma^*}(y)\begin{cases}\max \sum_{j\notin\sigma^*}(c_k-\pi_{\sigma^*}A_k)\,x_k\\ \text{s.t. } A_{\sigma^*}v+\sum_{k\notin\sigma^*}A_kx_k=b\\ v\in\mathbb{R},\, x_k\in\mathbb{R},\, 0\le x_k\le 1\quad \forall k\notin\sigma^*\end{cases}\tag{6.24}$$

is the exact continuous analogue of the group relaxation $\mathrm{IP}_{\sigma^*}(y)$. Its optimal value is zero because the cost has negative coefficients, and its optimal solution is the optimal solution $x^*\in\mathbb{R}^n_+$ of **P** (take $v=A_{\sigma^*}^{-1}b\ge 0$ and $x_k^*=0$ for all $k\notin\sigma^*$). Thus,

$$c'x(\sigma^*)+\max\mathrm{LP}_{\sigma^*}(y)=c'x(\sigma^*)=f(y,c),$$

exactly as (6.19) for **P** in Theorem 6.1.

Moreover, for a feasible basis $\sigma\neq\sigma^*$ of the linear program **P**, the LP

$$\mathrm{LP}_{\sigma}(y)\begin{cases}\max \sum_{k\in M_\sigma^+}(c_k-\pi_\sigma A_k)\,(x_k-1)+\sum_{k\in M_\sigma^-}(c_k-\pi_\sigma A_k)\,x_k\\ \text{s.t. } A_\sigma v+\sum_{k\notin\sigma}A_kx_k=y,\\ v\in\mathbb{R},\, x_k\in\mathbb{R},\, 0\le x_k\le 1\quad \forall k\notin\sigma\end{cases}\tag{6.25}$$

has also optimal value zero; take $x_k=1$ (resp., $x_k=0$) for $k\in M_\sigma^+$ (resp., $k\in M_\sigma^-$) because $c_k-\pi_\sigma A_k>0$ (resp., $c_k-\pi_\sigma A_k<0$) whenever $k\in M_\sigma^-$ (resp., $k\in M_\sigma^+$). Hence,

$$c'x(\sigma)+\max\mathrm{LP}_\sigma(y)=c'x(\sigma)\;<\;f(y,c).$$

Therefore, uniqueness of the optimal basis $\sigma^*\in\mathscr{B}(\Delta,\gamma)$ (when c is regular, and the optimal vertex x^* is nondegenerate) is equivalent to the uniqueness of the max in

$$\max_{x(\sigma):\,\sigma\in\mathscr{B}(\Delta,\gamma)}\left\{c'x(\sigma)+\max\mathrm{LP}_\sigma(y)\right\}\tag{6.26}$$

(with $\mathrm{LP}_\sigma(y)$ as in (6.25)), which is also attained at the unique basis σ^*.

Therefore, it makes sense to state the following.

Definition 6.1. Let $c\in\mathbb{R}^n$ be regular and $y\in\Lambda\cap\overline{\gamma}$. The integer program $\mathbf{P}_d$ has a unique optimal basis if the max in (5.3), or, equivalently, the max in

$$\max_{x(\sigma):\,\sigma\in\mathscr{B}(\Delta,\gamma)}\left\{c'x(\sigma)+\max\mathrm{IP}_\sigma(y)\right\}\tag{6.27}$$

is attained at a unique basis $\sigma^*\in\mathscr{B}(\Delta,\gamma)$ (in which case σ^* is also an optimal basis of the linear program **P**).

Note that, when c is regular and $y\in\gamma$ (so that $\Omega(y)$ is a simple polyhedron), the linear program **P** has a unique optimal basis σ^* which may not be an optimal basis of the integer program $\mathbf{P}_d$, i.e., the max in (6.27) is not attained at a unique σ (see Example 6.1). In other words, the *nondegeneracy property* for integer programming is a stronger condition than the nondegeneracy property in linear programming.

To see what happens in the case of multiple maximizers σ in (6.27), consider the following elementary knapsack example.

Example 6.1. Let $A := [2,7,1] \in \mathbb{Z}^{1\times 3}$, $c := (5,17,1) \in \mathbb{N}^3$, $y := 5 \in \Lambda$. There is a unique chamber $\gamma = (0,+\infty)$ and the optimal value $f_d(y,c)$ is 11 with optimal solution $x^* = (2,0,1)$. However, with $\sigma^* = \{1\}$, $A_{\sigma^*} = [2]$, the group relaxation

$$\mathrm{IP}_{\sigma^*}(y) \to \begin{cases} \max(17-35/2)x_2 + (1-5/2)x_3, \\ 2v + 7x_2 + x_3 = 5, \\ v \in \mathbb{Z},\, x_2, x_3 \in \mathbb{N},\, x_2, x_3 < 2 \end{cases}$$

has optimal value $-1/2$ at $x_1 = -1, x_2 = 1, x_3 = 0$. Therefore, $c'x(\sigma^*) - 1/2 = 5 \times 5/2 - 1/2 = 12$. On the other hand, let $\sigma := \{2\}$ with $A_\sigma = [7]$. The group relaxation

$$\mathrm{IP}_{\sigma}(y) \to \begin{cases} \max(5-34/7)(x_1-7) + (1-17/7)x_3, \\ 2x_1 + 7v + x_3 = 5, \\ v \in \mathbb{Z},\, x_1, x_3 \in \mathbb{N},\, x_1, x_3 < 7 \end{cases}$$

has optimal value $-1/7$ at $x_1 = 6, x_3 = 0, v = -1$, and thus $c'x(\sigma) - 1/7 = 17 \times 5/7 - 1/7 = 84/7 = 12$. In Lemma 6.1(b), as $r \to \infty$, we have

$$\mathrm{e}^{c'x(\sigma^*)} U_{\sigma^*}(y, rc) \approx u^{12q} \text{ and } \mathrm{e}^{c'x(\sigma)} U_{\sigma}(y, rc) \approx -u^{12q},$$

and in fact, these two terms have the same coefficient but with opposite signs and thus cancel in the evaluation of $\lim_{r\to\infty} \widehat{f_d}(y, rc)^{1/r}$ in (1.4).

Thus, in this case, $\lim_{r\to\infty} \widehat{f_d}(y, rc)^{1/r}$ is not provided by the leading term of $\mathrm{e}^{rc'x(\sigma^*)} U_{\sigma^*}(y, rc)$ as a function of $u = \mathrm{e}^r$. We need to examine smaller powers of $P_{\sigma y}(u)$ for all σ.

Had we $c_2 = 16$ instead of $c_2 = 17$, then $\mathrm{IP}_{\sigma^*}(y)$ and $\mathrm{IP}_\sigma(y)$ would have $-3/2$ and $-3/7$ as respective optimal values, and with the same optimal solutions as before. Thus, again,

$$c'x(\sigma^*) - 3/2 = (25-3)/2 = 11, \quad c'x(\sigma) - 3/7 = (80-3)/7 = 11$$

have same value 11, which is now the optimal value $f_d(y,c)$. But first observe that the optimal solution x^* of $\mathbf{P}_d$ is also an optimal solution of $\mathrm{IP}_{\sigma^*}(y)$. Moreover, this time

$$\frac{1}{\mu(\sigma^*)} \mathrm{e}^{rc'x(\sigma^*)} U_{\sigma^*}(y, rc) \approx 2u^{11q},$$

because the integer program $\mathrm{IP}_{\sigma^*}(y)$ has two optimal solutions. (See (6.64) in Section 6.6 and (6.68) in Section 6.6 for the respective coefficients of the leading monomials of $P_{\sigma y}(u)$ and $Q_{\sigma y}(u)$.)

On the other hand,

$$\frac{1}{\mu(\sigma)} \mathrm{e}^{rc'x(\sigma)} U_{\sigma}(y, rc) \approx -u^{11q}.$$

Therefore, both $c'x(\sigma^*) + \max \mathrm{IP}_{\sigma^*}(y)$ *and* $c'x(\sigma) + \max \mathrm{IP}_\sigma(y)$ provide the optimal value $f_d(y,c)$ in (5.3). In this example, the uniqueness uniqueness property in Definition 6.1 does *not* hold because the max in (6.27) is *not* attained at a unique σ. However, note that the linear program $\mathbf{P}$ has a unique optimal basis σ^*.

We have seen that the optimal value $f_d(y,c)$ may not be provided by the group relaxation $\mathrm{IP}_{\sigma^*}(y)$ when the max in (6.27) (or (5.3)) is attained at several bases $\sigma \in \mathscr{B}(\Delta,\gamma)$ (let Γ be the set of such bases). This is because, as a function of $\mathrm{e}^{r/q}$, the leading monomials of $\mu(\sigma)^{-1}\mathrm{e}^{c'x(\sigma)}U_\sigma(y,rc)$, $\sigma \in \Gamma$, have coefficients with different signs which permit their possible cancellation in the evaluation of $\lim_{r\to\infty} \widehat{f}_d(y,rc)$ in (1.4). The coefficient of the leading monomial of $P_{\sigma y}$ is positive ($=\mu(\sigma)$), whereas the coefficient of the leading monomial of $Q_{\sigma y}$ is given by $(-1)^{a_\sigma}$, where $a_\sigma = |M_\sigma^+|$ (see Section 6.6). Thus, if we list the vertices of the linear program $\mathbf{P}$ in decreasing order according to their value, the second, fourth, ..., vertices are the only ones capable of a possible cancellation, because their corresponding leading monomials have a negative coefficient.

6.4 The Knapsack Problem

We here consider the so-called knapsack problem, that is, when $m=1$, $A \in \mathbb{N}^{1\times n} = \{a_j\}$, $c \in \mathbb{Q}^n$, $y \in \mathbb{N}$. In this case, with $s := \sum_j a_j$, the generating function $\widehat{F}_d$ reads

$$z \mapsto \widehat{F}_d(z,c) = \frac{1}{\prod_{j=1}^n (1-\mathrm{e}^{c_j} z^{-a_j})} = \frac{z^s}{\prod_{j=1}^n (z^{a_j} - \mathrm{e}^{c_j})}, \tag{6.28}$$

which is well-defined provided $|z|^{a_j} > c_j$ for all $j = 1,\dots,n$. After possible multiplication by an integer, we may and will assume that $c \in \mathbb{N}^n$. If $c \in \mathbb{N}^n$ is regular then, after relabeling if necessary, we have

$$c_1/a_1 > c_2/a_2 > \cdots > c_n/a_n. \tag{6.29}$$

So with $r \in \mathbb{N}$, letting $u := \mathrm{e}^r$, the function

$$\frac{\widehat{F}_d(z,rc)}{z} = \frac{z^{s-1}}{\prod_{j=1}^n (z^{a_j} - u^{c_j})} \tag{6.30}$$

may be decomposed with respect to z into simpler rational fractions of the form

$$\frac{\widehat{F}_d(z,rc)}{z} = \sum_{j=1}^n \frac{P_j(u,z)}{(z^{a_j} - u^{c_j})}, \tag{6.31}$$

where $P_j(u,\cdot) \in \mathbb{R}[z]$ has degree at most $a_j - 1$, and $P_j(\cdot,z)$ is a rational fraction of u. This decomposition can be obtained by symbolic computation.

Next, write

$$P_j(u,z) = \sum_{k=0}^{a_j-1} P_{jk}(u) z^k \qquad j = 1,\dots,n, \tag{6.32}$$

where the $P_{jk}(u)$ are rational fractions of u, and let $\rho > rc_1/a_1$. We then have

$$\begin{aligned}\widehat{f_d}(y,rc) &= \sum_{j=1}^{n}\sum_{k=0}^{a_j-1} P_{jk}(\mathrm{e}^r)\int_{|z|=\rho}\frac{z^{y+k}}{(z^{a_j}-\mathrm{e}^{rc_j})}\,dz \\ &= \sum_{j=1}^{n}\sum_{k=0}^{a_j-1} P_{jk}(\mathrm{e}^r)\begin{cases}\mathrm{e}^{rc_j(y+k+1-a_j)/a_j} & \text{if } y+k+1=0(\bmod a_j)\\ 0 & \text{otherwise.}\end{cases}\end{aligned} \tag{6.33}$$

Equivalently, letting $y = s_j(\bmod a_j)$ for all $j = 1,\ldots,n$,

$$\widehat{f_d}(y,rc) = \sum_{j=1}^{n} P_{j(a_j-s_j-1)}(\mathrm{e}^r)\mathrm{e}^{rc_j(y-s_j)/a_j}, \tag{6.34}$$

so that

$$f_d(y,c) = \lim_{r\to\infty}\frac{1}{r}\ln\left[\sum_{j=1}^{n} P_{j(a_j-s_j-1)}(\mathrm{e}^r)\mathrm{e}^{rc_j(y-s_j)/a_j}\right], \tag{6.35}$$

and if the max in (5.3) (or (6.27)) is attained at a unique basis σ^*, then $\sigma^* = \{1\}$, and

$$f_d(y,c) = c_1(y-s_1)/a_1 + \lim_{r\to\infty}\frac{1}{r}\ln P_{1(a_1-s_1-1)}(\mathrm{e}^r). \tag{6.36}$$

So, if one has computed symbolically the functions $P_{jk}(u)$, it suffices to *read* the power of the leading term of $P_{1(a_1-s_1-1)}(u)$ as $u\to\infty$ to obtain $f_d(y,c)$ by (6.36).

Example 6.2. Let us go back to Example 6.1 with $A = [2,7,1]$, $c = [5,17,1]$, and $y = 5$. The optimal basis of the continuous knapsack is $\sigma = \{1\}$ with $A_\sigma = [2]$. Symbolic computation of $P_1(u,z)$ gives

$$P_1(u,z) = \frac{u^6+u^5}{(u^3-1)(u-1)} + \frac{u^4+u}{(u^3-1)(u-1)}z,$$

and therefore, as $s_1 = 1$,

$$P_{10}(u) = \frac{u^6+u^5}{(u^3-1)(u-1)}, \tag{6.37}$$

with leading term $u^{6-4} = u^2$, so that with $y = 5$,

$$\frac{c_1(y-s_1)}{a_1} + \lim_{r\to\infty}\frac{1}{r}\ln P_{1(a_1-s_1-1)}(\mathrm{e}^r) = \frac{5(5-1)}{2} + 6 - 4 = 12.$$

Similarly, with $\sigma = \{2\}$, $A_\sigma = [7]$, $s_2 = 5$, the term $P_{2(7-5-1)}(u)$ is

$$-\frac{u^{14}+u^{15}+u^{17}+u^{18}+u^{20}+u^{21}+u^{23}}{(u^{10}-1)(u-1)}, \tag{6.38}$$

with leading term $-u^{23-11} = -u^{12}$, so that with $y = 5$,

$$\frac{c_2(y-s_2)}{a_2} + \lim_{r\to\infty}\frac{1}{r}\ln P_{1(a_1-s_1-1)}(\mathrm{e}^r) = 0 + 23 - 11 = 12,$$

and we retrieve that the max in (6.27) is not unique. Moreover, the two leading terms u^{12} and $-u^{12}$ cancel in (6.35). On the other hand, the next leading term in (6.37) is now $u^{5-4} = u$, whereas the next leading term in (6.38) is now $-u^{21-11} = -u^{10}$, and the optimal value 11 of $\mathbf{P}$ is provided by $10+$ the power of the next leading term in (6.37), i.e., $10+1=11$.

Nesterov's algorithm for knapsack

From the simple observation that the coefficients of products of polynomials can be computed efficiently by fast Fourier transform (FFT), Nesterov [114] proposed applying this technique to generate the first y coefficients of the expansion of $\widehat{F}_d(z,rc)$ for r sufficiently large. Indeed, with $r > n\ln(1+y)^{-1}$,

$$-1+\frac{1}{r}\ln\widehat{f}_d(y,rc) < f_d(y,c) \leq \frac{1}{r}\ln\widehat{f}_d(y,rc).$$

The algorithm consists of two steps. The first step computes the coefficients of the univariate polynomial $z \mapsto p(z) := \widehat{F}_d(z,rc)^{-1}$. The second computes the first $y+1$ coefficients of the expansion of the rational function $z \mapsto p(z)^{-1} = \widehat{F}_d(z,rc)$. And we get this:

Theorem 6.2. *With $a \in \mathbb{N}^n, y \in \mathbb{N}$, and $c \in \mathbb{R}^n$, consider the unbounded knapsack problem $f_d(y,c) := \max\{c'x \,:\, a'x = y;\, x \in \mathbb{N}^n\}$. Then $f_d(y,c)$ can be computed in $O(\|a\|_1\ln(\|a\|_1)\ln n + (\ln n)^2 y)$ operations of exact real arithmetics (where $\|a\|_1 = \sum_{i=1}^n |a_i|$).*

The next interesting question is Can we provide an explicit description of what would be a *dual* of the integer program $\mathbf{P}_d$? The purpose of the next section is to present such a dual problem $\mathbf{P}_d^*$.

6.5 A dual of $\mathbf{P}_d$

In this section we provide a formulation of a problem $\mathbf{P}_d^*$, a dual of $\mathbf{P}_d$, which is an analogue of the LP dual $\mathbf{P}^*$ of the linear program $\mathbf{P}$.

Recall that the value function $y \mapsto f(y,c)$ of the linear program $\mathbf{P}$ is *concave*, and one has the well-known *convex duality* result

$$f(y,c) = \inf_{\lambda\in\mathbb{R}^m} y'\lambda - f^*(\lambda,c), \tag{6.39}$$

where $f^*(\cdot,c) : \mathbb{R}^m \to \mathbb{R}\cup\{\infty\}$ given by

$$\lambda \mapsto f^*(\lambda,c) := \inf_{y\in\mathbb{R}^m} y'\lambda - f(y,c) = \begin{cases} 0 & \text{if } A'\lambda - c \geq 0 \\ -\infty & \text{otherwise} \end{cases}$$

is the Legendre–Fenchel transform of $f(\cdot,c)$ (in the concave case). In addition,

$$f(y,c) = \inf_{\lambda \in \mathbb{R}^m} y'\lambda - f^*(\lambda, c) = \min_{\lambda \in \mathbb{R}^m} \{ y'\lambda \,|A'\lambda \geq c \} \tag{6.40}$$

and we retrieve in (6.40) the usual dual $\mathbf{P}^*$ of the linear program $\mathbf{P}$. But we can also write (6.39) as

$$f(y,c) = \inf_{\lambda \in \mathbb{R}^m} \sup_{x \in \mathbb{R}^n_+} \lambda'(y - Ax) + c'x, \tag{6.41}$$

or, equivalently,

$$\mathrm{e}^{f(y,c)} = \inf_{\lambda \in \mathbb{R}^m} \sup_{x \in \mathbb{R}^n_+} \mathrm{e}^{\lambda'(y-Ax)} \mathrm{e}^{c'x}. \tag{6.42}$$

Define now the following optimization problem:

$$\mathbf{P}_d^*: \quad \gamma^*(y) := \inf_{z \in \mathbb{C}^m} \sup_{x \in \mathbb{N}^n} \Re\left(z^{y-Ax} \mathrm{e}^{c'x}\right) = \inf_{z \in \mathbb{C}^m} f_c^*(z,y), \tag{6.43}$$

where $u \mapsto \Re(u)$ denotes the real part of $u \in \mathbb{C}$.

Clearly, the function $f_c^*(\cdot, y) : \mathbb{C}^m \to \mathbb{R}$,

$$z \mapsto f_c^*(z,y) = \sup_{x \in \mathbb{N}^n} \Re\left(z^{y-Ax} \mathrm{e}^{c'x}\right)$$

$$= \sup_{x \in \mathbb{N}^n} \Re\left(z^y \prod_{j=1}^n \left(z^{-A_j} \mathrm{e}^{c_j}\right)^{x_j}\right) \tag{6.44}$$

is finite if and only if $|z^{A_j}| \geq \mathrm{e}^{c_j}$ for all $j = 1, \ldots, n$, or equivalently, if and only if $A' \ln |z| \geq c$. That is, $\ln |z|$ should be in the feasible set of the dual linear program $\mathbf{P}^*$ in (6.40).

We claim that $\mathbf{P}_d^*$ is a dual problem of $\mathbf{P}_d$. Indeed, under an appropriate rescaling $c \to \tilde{c} := \alpha c$ of the cost vector c, and a condition on the group $G(\sigma^*)$ associated with the optimal basis σ^* of the linear program $\mathbf{P}$, both problems $\mathbf{P}_d^*$ and $\mathbf{P}_d$ have the same optimal value.

Theorem 6.3. *Let $y \in \Lambda \cap \overline{\gamma}$ and $c \in \mathbb{R}^n$ be regular. Assume that the integer program $\mathbf{P}_d$ has a feasible solution, and the uniqueness property holds (see Definition 6.1). Let $\sigma^* \in \mathscr{B}(\Delta, \gamma)$ be an optimal basis of the linear program $\mathbf{P}$ and let λ^* be the corresponding optimal solution of the dual linear program $\mathbf{P}^*$ in (6.40).*

Assume that there exists $g^* \in G(\sigma^*)$ such that $\mathrm{e}^{2i\pi u}(g^*) \neq 1$ whenever $u \notin \oplus_{j \in \sigma^*} A_j \mathbb{Z}$. Let $\tilde{c} := \alpha c$ with $\alpha > 0$. If α is sufficiently small, i.e., $0 < \alpha < \overline{\alpha}$ for some $\overline{\alpha} \in \mathbb{R}^+$, then

(i) The optimal value $f_d(y,c)$ of $\mathbf{P}_d$ satisfies

$$\mathrm{e}^{\alpha f_d(y,c)} = \mathrm{e}^{f_d(y,\tilde{c})} = \gamma^*(y) = \inf_{z \in \mathbb{C}^m} \sup_{x \in \mathbb{N}^n} \Re\left(z^{y-Ax} \mathrm{e}^{\tilde{c}'x}\right) = \inf_{z \in \mathbb{C}^m} f_{\tilde{c}}^*(z,y) \tag{6.45}$$

(ii) For $z_{g^*} \in \mathbb{C}^m$ as in (5.12) (with $\alpha\lambda^*$ in lieu of λ^*),

$$e^{\alpha f_d(y,c)} = e^{f_d(y,\tilde{c})} = \gamma^*(y) = \sup_{x\in\mathbb{N}^n} \Re\left(z_{g^*}^{y-Ax} e^{\tilde{c}'x}\right) = f_{\tilde{c}}^*(z_{g^*}, y) \tag{6.46}$$

For a proof see Section 6.6. The assumption on g^* in Theorem 6.3 is satisfied in particular when the group $G(\sigma^*)$ is cyclic (which for instance is the case for the knapsack problem). In view of the uniqueness property required to hold in Theorem 6.3, $\mathbf{P}_d^*$ is also a dual of the Gomory asymptotic group relaxation $\mathrm{IP}_{\sigma^*}(y)$ in (6.2), as the latter is exact.

Notice that $\mathbf{P}_d^*$ becomes the usual linear program $\mathbf{P}^*$ of $\mathbf{P}$ when $z \in \mathbb{C}^m$ is replaced with $|z| \in \mathbb{R}^m$. Indeed, from (6.39), we have

$$\begin{aligned} f(y,c) &= \inf_{\lambda\in\mathbb{R}^m} \sup_{x\in\mathbb{R}^n_+} \lambda'(y-Ax) + c'x = \inf_{\lambda\in\mathbb{R}^m} \lambda'y + \sup_{x\in\mathbb{R}^n_+} (c'-A'\lambda)x \\ &= \inf_{\lambda\in\mathbb{R}^m} \lambda'y + \sup_{x\in\mathbb{N}^n} (c'-A'\lambda)x = \inf_{\lambda\in\mathbb{R}^m} \sup_{x\in\mathbb{N}^n} \lambda'(y-Ax) + c'x \end{aligned}$$

that is, one may replace the sup over $\mathbb{R}^m_+$ by the sup over $\mathbb{N}^n$. Therefore, (6.42) becomes

$$e^{f(y,c)} = \inf_{\lambda\in\mathbb{R}^m} \sup_{x\in\mathbb{N}^n} e^{\lambda'(y-Ax)} e^{c'x}. \tag{6.47}$$

On the other hand, if in (6.43), we replace $z \in \mathbb{C}^m$ by the vector of its component moduli $|z| = e^{\lambda} \in \mathbb{R}^m$, we obtain

$$\inf_{z\in\mathbb{C}^m} \sup_{x\in\mathbb{N}^n} \Re\left(|z|^{y-Ax} e^{c'x}\right) = \inf_{\lambda\in\mathbb{R}^m} \sup_{x\in\mathbb{N}^n} e^{\lambda'(y-Ax)} e^{c'x} = e^{f(y,c)}. \tag{6.48}$$

Hence, when the uniqueness property (see Definition 6.1) holds for $\mathbf{P}_d$, Table 5.2 can be completed by

$$f(y,c) = \inf_{z\in\mathbb{C}^m} \sup_{x\in\mathbb{N}^n} \Re\left(|z|^{y-Ax} e^{c'x}\right) = \inf_{z\in\mathbb{C}^m} f_c^*(|z|, y),$$

$$f_d(y,c) = \inf_{z\in\mathbb{C}^m} \sup_{x\in\mathbb{N}^n} \Re\left(z^{y-Ax} e^{c'x}\right) = \inf_{z\in\mathbb{C}^m} f_c^*(z, y).$$

Again, as for the vertex residue function (5.13), a complete analogy between $\mathbf{P}_d^*$ and the dual linear program $\mathbf{P}^*$ in (6.40) (equivalently (6.47)) is obtained by just a change of $z \in \mathbb{C}^m$ with $|z| \in \mathbb{R}^m$. (Compare with Table 5.2 in Chapter 5.)

6.6 Proofs

We first state two results that we use in the proofs of Lemma 6.1, Corollary 6.1, and Theorem 6.3.

Auxiliary result

Let $e : \mathbb{R} \to \mathbb{C}$ be the function $x \mapsto e(x) := \mathrm{e}^{2i\pi x}$. First note that for all $m \in \mathbb{Z}$, $s \in \mathbb{N}$, we have the identity

$$\sum_{k=1}^{s} e(mk/s) = \begin{cases} s & \text{if } m = 0(\text{mod } s) \\ 0 & \text{otherwise} \end{cases} \tag{6.49}$$

But we also have the following result:

Lemma 6.2. *Let $m \in \mathbb{N}$ and $\{z_j\}_{j=1,\ldots,m} \subset \mathbb{C}$ be the roots of $z^m - 1 = 0$. Then for all $k = 1, \ldots, m-1$*

$$\begin{aligned} \sum_{\substack{1 \le i_1 < i_2 \cdots < i_k \le m \\ i_1, \ldots, i_k \ne j}} z_{i_1} \ldots z_{i_k} &= \sum_{\substack{1 \le i_1 < i_2 \cdots < i_k \le m \\ i_1, \ldots, i_k \ne j}} e\left(\frac{(i_1 + \cdots + i_k)}{m}\right) \\ &= (-1)^k z_j^k = (-1)^k e\left(\frac{kj}{m}\right). \end{aligned} \tag{6.50}$$

Proof. The proof is by induction. For $k = 1$ we have

$$\sum_{\substack{1 \le i \le m \\ i \ne j}} z_i = \sum_{i=1}^{m} z_i - z_j = 0 - z_j,$$

because the z_j are roots of $z^m - 1 = 0$. Next, assume that (6.50) holds for $l = 1, \ldots, k$. Then, as the z_j are roots of $z^m - 1 = 0$, we have

$$\sum_{1 \le i_1 < i_2 \cdots < i_{k+1} \le m} z_{i_1} \ldots z_{i_{k+1}} = 0.$$

Hence,

$$\begin{aligned} \sum_{\substack{1 \le i_1 < i_2 \cdots < i_{k+1} \le m \\ i_1, \ldots, i_{k+1} \ne j}} z_{i_1} \ldots z_{i_{k+1}} &= \sum_{1 \le i_1 < i_2 \cdots < i_{k+1} \le m} z_{i_1} \ldots z_{i_{k+1}} - \sum_{\substack{1 \le i_1 \cdots i_k \le m \\ i_1, \ldots, i_k \ne j}} z_j z_{i_1} \ldots z_{i_k} \\ &= 0 - z_j \times \sum_{\substack{1 \le i_1 \cdots i_k \le m \\ i_1, \ldots, i_k \ne j}} z_{i_1} \ldots z_{i_k} \\ &= -z_j (-1)^k z_j^k \quad \text{(by the induction hypothesis)} \\ &= (-1)^{k+1} z_j^{k+1}. \end{aligned}$$

Some properties of the characters of $G(\sigma)$

With $\sigma \in \mathscr{B}(\Delta)$, let $G(\sigma)$ be the group $(\oplus_{j\in\sigma}A_j\mathbb{Z})^*/\Lambda^*$, of order $\mu(\sigma) =: s$, and write

$$G(\sigma) = \{g_1,\dots,g_s\}.$$

Let $y \in \Lambda$ with $y \notin \oplus_{j\in\sigma}A_j\mathbb{Z}$ and consider the character $e^{2i\pi y}$ of $G(\sigma)$. Then $e^{2i\pi y}(g) = e(-g'A_\sigma^{-1}y) = e(v_g/s)$ for some $v_g \in \mathbb{N}, v_g < s$. That is, the mapping $g \mapsto e^{2i\pi y}(g)$ sends the group $G(\sigma)$ into a subgroup of the multiplicative group of the s roots of unity. Let $s_y < s$ (with $s = p_y s_y$ for some $p_y \in \mathbb{N}$) be the order of this subgroup (which consists of the roots $\{e(j/s_y)\}$, $j = 1,\dots,s_y$). Equivalently, s_y is the smallest integer such that $ys_y \in \oplus_{j\in\sigma}A_j\mathbb{Z}$. We can define a partition of $G(\sigma)$ (which depends on y) into s_y equivalence classes $\{C_i^y\}_{i=1}^{s_y}$ of same cardinal $p_y := s/s_y$ by setting

$$g \sim g' \quad \Leftrightarrow \quad e^{2i\pi y}(g) = e^{2i\pi y}(g'), \quad g,g' \in G(\sigma). \tag{6.51}$$

We next denote by $\tilde{g}_i$ a representative of the class C_i^y and by $G_y(\sigma)$ the set $\{C_1^y,\dots,C_{s_y}^y\}$ of equivalence classes. We have the following result:

Lemma 6.3. *Let $y \in \Lambda$ with $y \notin \oplus_{j\in\sigma}A_j\mathbb{Z}$ and let $\{C_i^y\}$ be the equivalence classes defined by (6.51). Then*

(i) for all $j \in \mathbb{N}$ with $j < s_y$, we have

$$\sum_{1\le i_1<i_2,\cdots<i_j\le s_y} e^{2i\pi y}(\tilde{g}_{i_1}+\cdots+\tilde{g}_{i_j}) = 0. \tag{6.52}$$

(ii) For all $q \in \{1,\dots,s_y\}$ and $j < s_y$,

$$\sum_{\substack{1\le i_1<i_2<\cdots<i_j\le s_y \\ i_1,\dots,i_j\neq q}} e^{2i\pi y}(\tilde{g}_{i_1}+\cdots+\tilde{g}_{i_j}) = (-1)^j e^{2i\pi yj}(\tilde{g}_q). \tag{6.53}$$

Proof. As $e^{2i\pi y}(\tilde{g}_i) = e(i/s_y)$ for all $i = 1,\dots,s_y$, (i) and (ii) follow from Lemma 6.2.

We also have

Lemma 6.4. *Let $y \in \Lambda$. Then*

$$\sum_{g\in G(\sigma)} e^{2i\pi y}(g) = \begin{cases} \mu(\sigma) & \text{if } y \in \oplus_{j\in\sigma}A_j\mathbb{Z} \\ 0 & \text{otherwise} \end{cases} \tag{6.54}$$

and

$$\prod_{\tilde{g}\in G_y(\sigma)} e^{2i\pi y}(\tilde{g}) = (-1)^{s_y+1}. \tag{6.55}$$

Proof. If $y \in \oplus_{j\in\sigma}A_j\mathbb{Z}$ then $e^{2i\pi y}(g) = 1$ for all $g \in G(\sigma)$, which yields the first part of (6.54). On the other hand, if $y \notin \oplus_{j\in\sigma}A_j\mathbb{Z}$ we proceed as before. Let $G_y(\sigma)$ be the set of s_y equivalence classes of $G(\sigma)$ defined in (6.51). We thus have

$$\sum_{g\in G(\sigma)} \mathrm{e}^{2i\pi y}(g) = p_y \sum_{\tilde{g}\in G_y(\sigma)} \mathrm{e}^{2i\pi y}(\tilde{g}) = p_y \sum_{j=1}^{s_y} e(j/s_y) = 0,$$

which proves (6.54). Next,

$$\prod_{\tilde{g}\in G_y(\sigma)} \mathrm{e}^{2i\pi y}(\tilde{g}) = \prod_{j=1}^{s_y} e(j/s_y) = (-1)^{s_y+1},$$

which proves (6.55).

Proof of Lemma 6.1

From (6.7) we have

$$U_\sigma(y,rc) = \sum_{g\in G(\sigma)} \frac{\mathrm{e}^{2i\pi y}(g)}{\prod_{k\notin\sigma}(1-\mathrm{e}^{-2i\pi A_k}(g)\mathrm{e}^{r(c_k-\pi_\sigma A_k)})}, \tag{6.56}$$

and so letting $u := \mathrm{e}^{r/q}$ we have

$$U_\sigma(y,rc) = \sum_{g\in G(\sigma)} \frac{\mathrm{e}^{2i\pi y}(g)}{\prod_{k\notin\sigma}(1-\mathrm{e}^{-2i\pi A_k}(g)u^{q(c_k-\pi_\sigma A_k)})}. \tag{6.57}$$

Let S_σ be as in (6.8). As $\mathrm{e}^{-2i\pi A_k}(g) = 1$ whenever $A_k \in \oplus_{j\in\sigma} A_j\mathbb{Z}$,

$$U_\sigma(y,rc) = \frac{1}{\prod_{k\notin\sigma\cup S_\sigma}(1-u^{q(c_k-\pi_\sigma A_k)})} \sum_{g\in G(\sigma)} \frac{\mathrm{e}^{2i\pi y}(g)}{\prod_{k\in S_\sigma}(1-\mathrm{e}^{-2i\pi A_k}(g)u^{q(c_k-\pi_\sigma A_k)})}, \tag{6.58}$$

which, after reduction to same denominator, can be written as

$$U_\sigma(y,rc) = \frac{P_{\sigma y}}{Q_{\sigma y}} \tag{6.59}$$

for two Laurent polynomials $P_{\sigma y}, Q_{\sigma y} \in \mathbb{R}[u,u^{-1}]$.

The Laurent polynomial $\mathbf{P}_{\sigma\mathbf{y}}(\mathbf{u})$

Write the finite group

$$G(\sigma) := (\oplus_{j\in\sigma} A_j\mathbb{Z})^*/\Lambda^*$$

as $G(\sigma) = \{g_1,\ldots,g_s\}$ with $s := \mu(\sigma)$, and for $k \in S_\sigma$, consider the character $\mathrm{e}^{-2i\pi A_k}$ of the group $G(\sigma)$. For $k \in S_\sigma$, define in $G(\sigma)$ the equivalence relationship

$$g \sim g' \Leftrightarrow \mathrm{e}^{-2i\pi A_k}(g) = \mathrm{e}^{-2i\pi A_k}(g'), \qquad g, g' \in G(\sigma).$$

According to $\sim$, one may partition $G(\sigma)$ into $s_{k\sigma}$ equivalence classes $\{C_i^k\}$ of the same cardinality $s/s_{k\sigma}$, where $s_{k\sigma} \in \mathbb{N}$ is the smallest integer for which $s_{k\sigma}A_k \in \oplus_{j\in\sigma}A_j\mathbb{Z}$ (see Section 6.6 for more details). Let $G_k(\sigma)$ be the set of $s_{k\sigma}$ equivalence classes of $G(\sigma)$. A representative of the equivalence class of $g \in G(\sigma)$ is denoted by $\tilde{g}$, so that $G_k(\sigma) := \{\tilde{g}_1, \ldots, \tilde{g}_{s_{k\sigma}}\}$. As $\mathrm{e}^{-2i\pi A_k}(g) = \mathrm{e}^{-2i\pi A_k}(g') = \mathrm{e}^{-2i\pi A_k}(\tilde{g})$ if $g \sim g'$,

$$P_{\sigma y}(u) = \sum_{g\in G(\sigma)} \mathrm{e}^{2i\pi y}(g) \prod_{k\in S_\sigma} \left[\prod_{\tilde{g}'\neq\tilde{g}} (1 - \mathrm{e}^{-2i\pi A_k}(\tilde{g}')u^{q(c_k - \pi_\sigma A_k)}) \right]. \tag{6.60}$$

Therefore, it follows that $P_{\sigma y}(u)$ is a sum of monomials of the form u^ν, $\nu \in \mathbb{Z}$, where

$$\nu = q \sum_{k\in S_\sigma} x_k(c_k - \pi_\sigma A_k) \quad \text{with } x_k \in \mathbb{N}, x_k < s_{k\sigma}, \quad k \in S_\sigma. \tag{6.61}$$

The corresponding coefficient of this monomial u^ν of $P_{\sigma y}(u)$ is given by

$$\sum_{g\in G(\sigma)} \mathrm{e}^{2i\pi y}(g) \prod_{k\notin\sigma} \Gamma(g,k),$$

with

$$\begin{aligned} \Gamma(g,k) &= \left\{ \begin{array}{c} (-1)^{x_k} \sum \mathrm{e}^{-2i\pi A_k}(\tilde{g}_{i_1} + \cdots + \tilde{g}_{i_{x_k}}) \\ \text{s.t. } 1 \leq i_1 < i_2 \cdots < i_{x_k} \leq s_{k\sigma} \\ \tilde{g}_{i_1}, \ldots, \tilde{g}_{i_{x_k}} \neq \tilde{g} \end{array} \right\} \\ &= \mathrm{e}^{-2i\pi A_k x_k}(\tilde{g}) = \mathrm{e}^{-2i\pi A_k x_k}(g) \end{aligned}$$

(by Lemma 6.3(b)).

Hence, the coefficient of the monomial u^ν of $P_{\sigma y}(u)$ with ν as in (6.61) is

$$\begin{aligned} \sum_{g\in G(\sigma)} \mathrm{e}^{2i\pi y}(g) \prod_{k\in S_\sigma} \Gamma(g,k) &= \sum_{g\in G(\sigma)} \mathrm{e}^{2i\pi y}(g) \prod_{k\in S_\sigma} \mathrm{e}^{-2i\pi A_k x_k}(g) \\ &= \sum_{g\in G(\sigma)} \mathrm{e}^{2i\pi(y - \sum_{k\in S_\sigma} A_k x_k)}(g) \\ &= \begin{cases} s & \text{if } (y - \sum_{k\in S_\sigma} A_k x_k) \in \oplus_{j\in\sigma} A_j\mathbb{Z}, \\ 0 & \text{otherwise} \end{cases} \end{aligned} \tag{6.62}$$

(see Lemma 6.4). Consequently, the maximum algebraic degree of $P_{\sigma y}$ is given by the leading monomial u^ν where

$$\nu = \left\{ \begin{array}{l} q \max \sum_{k\in S_\sigma} (c_k - \pi_\sigma A_k) x_k \\ \text{s.t. } A_\sigma x_\sigma + \sum_{k\in S_\sigma} A_k x_k = y, \\ \quad x_k \in \mathbb{N}, x_k < s_{k\sigma} \quad \forall k \in S_\sigma \\ \quad x_\sigma \in \mathbb{Z}^m, \end{array} \right. \tag{6.63}$$

and the coefficient of the monomial u^v of $P_{\sigma y}$ is

$$\mu(\sigma) \times \text{ the number of optimal solutions of (6.63).} \tag{6.64}$$

The Laurent polynomial $\mathbf{Q}_{\sigma \mathbf{y}}(\mathbf{u})$

For all $k \in S_\sigma$, let $s_{k\sigma}$, $G_k(\sigma)$ be as defined in Section 6.6. As $\mathrm{e}^{-2i\pi A_k}(g)$ is constant in each equivalence class, we may write the Laurent polynomial $Q_{\sigma y}$ as the product $Q^1_{\sigma y} Q^2_{\sigma y}$ of the two Laurent polynomials

$$Q^1_{\sigma y}(u) := \prod_{k \notin \sigma \cup S_\sigma} (1 - u^{q(c_k - \pi_\sigma A_k)}) \tag{6.65}$$

and

$$Q^2_{\sigma y}(u) := \prod_{k \in S_\sigma} \prod_{\tilde{g} \in G_k(\sigma)} (1 - \mathrm{e}^{-2i\pi A_k}(\tilde{g}) u^{q(c_k - \pi_\sigma A_k)}). \tag{6.66}$$

With arguments similar to those used for $P_{\sigma y}$ one may see that $Q^2_{\sigma y}$ is a Laurent polynomial with all powers of the form u^v, $v \in \mathbb{Z}$, where

$$v = q \sum_{k \in S_\sigma} (c_k - \pi_\sigma A_k) x_k, \quad x_k \in \mathbb{N}, \quad x_k \leq s_{k\sigma}.$$

And for the same reasons as for $P_{\sigma y}$, the only monomials u^v with nonzero coefficient are those for which

$$\sum_{k \in S_\sigma} A_k x_k \in \bigoplus_{j \in \sigma} A_j \mathbb{Z},$$

which is the case if $x_k = s_{k\sigma}$. So the maximum algebraic degree of $Q^2_{\sigma y}$ is obtained with $x_k = s_{k\sigma}$ for all $k \in S_\sigma$ with $(c_k - \pi_\sigma A_k) > 0$, and $x_k = 0$ otherwise. In addition, the coefficient of this monomial is

$$\prod_{k \in S_\sigma \cap M^+_\sigma} \left[(-1)^{s_{k\sigma}} \prod_{\tilde{g} \in G_k(\sigma)} \mathrm{e}^{-2i\pi A_k}(\tilde{g}) \right] = \prod_{k \in S_\sigma \cap M^+_\sigma} (-1)^{2s_{k\sigma}+1} = (-1)^{|S_\sigma \cap M^+_\sigma|},$$

where we have used Lemma 6.4.

Similarly, the maximum algebraic degree of $Q^1_{\sigma y}$ is given by the sum of $q(c_k - \pi_\sigma A_k)$ over all $k \notin \sigma \cup S_\sigma$ with $c_k - \pi_\sigma A_k > 0$. Therefore, the maximum algebraic degree of $Q_{\sigma y}$ is given by

$$\deg Q_{\sigma y} = q \sum_{\substack{k \notin \sigma \cup S_\sigma \\ k \in M^+_\sigma}} (c_k - \pi_\sigma A_k) + q s_{k\sigma} \sum_{\substack{k \in S_\sigma \\ k \in M^+_\sigma}} (c_k - \pi_\sigma A_k). \tag{6.67}$$

(In particular, it is zero for the optimal basis σ^* of the linear program **P**.) Finally, the coefficient of this leading monomial of $Q_{\sigma y}(u)$ is given by $(-1)^{a_\sigma}$ where

$$a_\sigma = |S_\sigma \cap M^+_\sigma| + \left| \{k \notin \sigma \cup S_\sigma;\, k \in M^+_\sigma\} \right| = |M^+_\sigma|. \tag{6.68}$$

This completes the proof of Lemma 6.1(i).
(ii) (6.13) just identifies the leading term when $r \to \infty$.

Proof of Corollary 6.1

In view of (6.63) and (6.67)

$$\frac{1}{q}[\deg P_{\sigma y} - \deg Q_{\sigma y}] = -\sum_{k \notin S_\sigma; k \in M_\sigma^+} (c_k - \pi_\sigma A_k)$$

$$+\begin{cases} \max \sum_{k \in S_\sigma \cap M_\sigma^+} (c_k - \pi_\sigma A_k)(x_k - s_{k\sigma}) + \sum_{k \in S_\sigma \cap M_\sigma^-} (c_k - \pi_\sigma A_k) x_k \\ \text{s.t.} \quad A_\sigma x_\sigma + \sum_{k \in S_\sigma} A_k x_k = y, \\ \quad x_\sigma \in \mathbb{Z}^m, x_k \in \mathbb{N}, x_k < s_{k\sigma} \quad \forall k \in S_\sigma. \end{cases} \tag{6.69}$$

Equivalently,

$$\frac{1}{q}[\deg P_{\sigma y} - \deg Q_{\sigma y}] =$$

$$\begin{cases} \max \sum_{k \in M_\sigma^+} (c_k - \pi_\sigma A_k)(x_k - s_{k\sigma}) + \sum_{k \in M_\sigma^-} (c_k - \pi_\sigma A_k) x_k \\ \text{s.t.} \quad A_\sigma x_\sigma + \sum_{k \in S_\sigma} A_k x_k = y, \\ \quad x_\sigma \in \mathbb{Z}^m, x_k \in \mathbb{N}, x_k < s_{k\sigma} \quad \forall k \in S_\sigma, \end{cases}$$

which is the integer program $\text{IP}_\sigma(y)$ in Corollary 6.1. This is because in the above integer program, obviously one should take

- $x_k = s_{k\sigma} - 1$ for all $k \notin S_\sigma, k \in M_\sigma^+$ and
- $x_k = 0$ for all $k \notin S_\sigma, k \in M_\sigma^-$,

which gives (6.69).

Proof of Theorem 6.3

First, notice that when c is replaced with αc (with $\alpha > 0$), then the optimal solutions of the integer program $\mathbf{P}_d$ and the linear programs $\mathbf{P}$ and $\mathbf{P}^*$ are the same, whereas their respective optimal values are rescaled by α. Next, observe that

$$f_d(y, \tilde{c}) \leq \ln \gamma^*(y) \leq f(y, \tilde{c}). \tag{6.70}$$

Indeed,

$$\begin{aligned}\inf_{z\in\mathbb{C}^m}\sup_{x\in\mathbb{N}^n}\Re\left(z^{y-Ax}\mathrm{e}^{\tilde{c}'x}\right) &\leq \inf_{z\in\mathbb{R}^m}\sup_{x\in\mathbb{R}^n_+}\Re\left(z^{y-Ax}\mathrm{e}^{\tilde{c}'x}\right)\\ &\leq \inf_{\lambda\in\mathbb{R}^m}\sup_{x\in\mathbb{R}^n_+}\mathrm{e}^{\lambda'(y-Ax)+\tilde{c}'x}\\ &= \mathrm{e}^{\inf_{\lambda\in\mathbb{R}^m}\sup_{x\in\mathbb{R}^n_+}\lambda'(y-Ax)+\tilde{c}'x}\\ &= \mathrm{e}^{f(y,\tilde{c})},\end{aligned}$$

which proves $\ln\gamma^*(y)\leq f(y,\tilde{c})$. Next, fix $z\in\mathbb{C}^m$ arbitrary, and let x^* be an optimal solution of $\mathbf{P}_d$. Then, as $(y-Ax^*)=0$, we obtain

$$\sup_{x\in\mathbb{N}^n}\Re\left(z^{y-Ax}\mathrm{e}^{\tilde{c}'x}\right)\geq \mathrm{e}^{\tilde{c}'x^*}=\mathrm{e}^{f_d(y,\tilde{c})},$$

and so, $\ln\gamma^*(y)\geq f_d(y,\tilde{c})$.

Now, let z_g be as in (5.12) with $\sigma=\sigma^*$ and $\alpha\lambda^*$ in lieu of λ^*. Recall that from (5.11), we have

$$z_g^{-A_j}\mathrm{e}^{\tilde{c}_j}=1\quad \forall j\in\sigma^*,\, g\in G(\sigma^*).$$

Therefore, for all $x\in\mathbb{N}^n$,

$$\begin{aligned}z_g^{y-Ax}\mathrm{e}^{\tilde{c}'x} &= z_g^{y-\sum_{k\notin\sigma^*}A_kx_k}\mathrm{e}^{\sum_{k\notin\sigma^*}\tilde{c}_kx_k}\prod_{j\in\sigma^*}\left(z_g^{-A_j}\mathrm{e}^{\tilde{c}_j}\right)^{x_j}\\ &= z_g^{y-\sum_{k\notin\sigma^*}A_kx_k}\mathrm{e}^{\sum_{k\notin\sigma^*}\tilde{c}_kx_k},\end{aligned}$$

and so,

$$\Re(z_g^{y-Ax}\mathrm{e}^{\tilde{c}'x})=\mathrm{e}^{\alpha y'\lambda^*}\,\mathrm{e}^{\sum_{k\notin\sigma^*}\alpha(c_k-A_k'\lambda^*)x_k}\times\cos\left[2\pi\theta_g'\left(y-\sum_{k\notin\sigma^*}A_kx_k\right)\right].$$

From this, we can deduce that

$$\begin{aligned}\sup_{x\in\mathbb{N}^n}\Re(z_g^{y-Ax}\mathrm{e}^{\tilde{c}'x})=\mathrm{e}^{\alpha y'\lambda^*}\sup_{x_k\in\mathbb{N},k\notin\sigma^*}&\left[\mathrm{e}^{\sum_{k\notin\sigma^*}\alpha(c_k-A_k'\lambda^*)x_k}\right.\\ &\left.\times\cos\left[2\pi\theta_g'\left(y-\sum_{k\notin\sigma^*}A_kx_k\right)\right]\right]\end{aligned}\tag{6.71}$$

Let x^* be an optimal solution of the group relaxation $\mathrm{IP}_{\sigma^*}(y)$. We claim that for $g=g^*$ (with g^* as in Theorem 6.3), the sup in (6.71) is attained at x^* with value the optimal value of $\mathbf{P}_d$. Suppose not. Then there exists $x_k\in\mathbb{N}$ for all $k\notin\sigma^*$ such that

$$\sum_{k\notin\sigma^*}(c_k-A_k'\lambda^*)x_k>\sum_{k\notin\sigma^*}(c_k-A_k'\lambda^*)x_k^* =: -\rho^*\tag{6.72}$$

(with $\rho^*>0$ because $c_k-A_k'\lambda^*<0,\, k\notin\sigma^*$), and

$$\cos\left[2\pi\theta'_{g^*}\left(y-\sum_{k\notin\sigma^*}A_kx_k\right)\right] > \mathrm{e}^{\sum_{k\notin\sigma^*}\alpha(c_k-A'_k\lambda^*)(x^*_k-x_k)}. \tag{6.73}$$

In addition, in view of our choice of g^*,

$$\cos\left[2\pi\theta'_{g^*}\left(y-\sum_{k\notin\sigma^*}A_kx_k\right)\right] < 1$$

because if $\cos[2\pi\theta'_{g^*}(y-\sum_{k\notin\sigma^*}A_kx_k)] = 1$ then $(y-\sum_{k\notin\sigma^*}A_kx_k) \in \oplus_{j\in\sigma^*}A_j\mathbb{Z}$, and x would be an admissible solution of the group relaxation $\mathrm{IP}_{\sigma^*}(y)$, in contradiction with the optimality of x^*.

Now, observe that $\cos[2\pi\theta'_{g^*}(y-\sum_{k\notin\sigma^*}A_kx_k)]$ takes finitely many values (and in fact, at most $\mu(\sigma^*)$ different values), because $(y-\sum_{k\notin\sigma^*}A_kx_k) \in \mathbb{Z}^m$. Thus,

$$1 > \delta := \max\{\cos(2\pi\theta'_{g^*}v) \,|\, v \in \mathbb{Z}^m,\ v \notin \oplus_{j\in\sigma^*}A_j\mathbb{Z}\}.$$

Moreover, from (6.72)

$$x_k \leq \frac{\sup_{k\notin\sigma^*}\rho^*}{(A'_k\lambda^*-c_k)} =: \beta, \qquad k\notin\sigma^*. \tag{6.74}$$

Hence,

$$1 > \delta > \mathrm{e}^{\sum_{k\notin\sigma^*}\alpha(c_k-A'_k\lambda^*)(x^*_k-x_k)}. \tag{6.75}$$

So, as x_k is bounded by β, one obtains a contradiction in (6.75) when α is sufficiently small. This proves (i) and (ii).

6.7 Notes

In this chapter we have made the link between Brion and Vergne's discrete formula and Gomory relaxations introduced by Gomory [58]. Most of the material of Section 6.2 is taken from Aardal et al. [1] and Gomory et al. [60]. Gomory relaxations can be qualified as a primal approach, as the relaxed integer program (6.2) is still defined in $\mathbb{Z}^n$ as $\mathbf{P}_d$. On the other hand, in practice, the master polyhedron is mainly used to generate cutting planes for $\mathbf{P}_d$ via subadditive functions, a dual approach. Reinforcements of Gomory relaxations where the nonnegativity constraints are relaxed only for a subset of the basic variables $\{x_j\}_{j\in\sigma}$ have been explored in, e.g., Thomas [128] and Wolsey [133, 135]. Theorem 6.2 as well as more details on the FFT algorithm for the knapsack problem can be found in Nesterov [114].

As already mentioned, there are other algebraic approaches to the integer program $\mathbf{P}_d$, using tools like *toric ideal*, *Gröbner fan*, and *Gröbner basis*. For instance, the Conti–Traverso algorithm [35] finds an optimal solution of $\mathbf{P}_d$ by first computing the reduced Gröbner basis G_c of a toric ideal related to $\mathbf{P}_d$ with respect to the cost vector c. Then with v any feasible solution of $\mathbf{P}_d$, one obtains an optimal solution v^* of $\mathbf{P}_d$ by computing the *normal form* x^{v^*} of x^v with respect to G_c. For more details, the interested reader is referred to the survey in Aardal et al. [1]. Interestingly, Gomory relaxations can also be reinterpreted in terms of those algebraic notions. Details can be found in, e.g., Hosten and Thomas [67].

Chapter 7
Barvinok's Counting Algorithm and Gomory Relaxations

7.1 Introduction

As already mentioned, solving the integer program $\mathbf{P}_d$ is still a formidable computational challenge. For instance, recall that the following small 5-variables integer program (taken from a list of hard knapsack problems in Aardal and Lenstra [2])

$$\begin{aligned}
&\min\{213x_1 - 1928x_2 - 11111x_3 - 2345x_4 + 9123x_5\} \\
&\text{s.t. } 12223x_1 + 12224x_2 + 36674x_3 + 61119x_4 + 85569x_5 = 89643482, \\
&\quad x_1, x_2, x_3, x_4, x_5 \in \mathbb{N}
\end{aligned} \tag{7.1}$$

is not solved after hours of computing with the last version of the CPLEX software package available in 2003, probably the best package at this time.

The first integer programming algorithm with polynomial time complexity when the dimension n is fixed is due to H.W. Lenstra [103] and uses lattice reduction technique along with a *rounding* of a convex body. As underlined in Barvinok and Pommersheim [15, p. 21], this rounding can be quite time consuming. Cook et al. [38] implemented a similar algorithm that uses the generalized basis reduction technique of Lovász and Scarf [108] and with no rounding of convex bodies.

On the other hand, Barvinok's counting algorithm described in Section 4.2, coupled with binary search, permits us to solve $\mathbf{I}_d$ (i.e., to evaluate $\widehat{f_d}(y,c)$) with polynomial time complexity when the dimension n is fixed. This algorithm has been successfully implemented in the software package `LattE` developed at the Mathematics Department of the University of California at Davis [40, 41].

In this chapter we describe two integer programming algorithms based on Barvinok's counting algorithm. The first one is Barvinok's counting algorithm coupled with a simple binary search and runs in time polynomial in the input size of the problem. The second has exponential computational complexity, but was found to be more efficient in some (limited) computational experiments for solving hard instances of knapsack problems taken from [2]. Finally, we relate Barvinok's counting formula with Gomory relaxations of integer programs and provide a simplified procedure for large values of y.

J.B. Lasserre, *Linear and Integer Programming vs Linear Integration and Counting*,
Springer Series in Operations Research and Financial Engineering, DOI 10.1007/978-0-387-09414-4_7,

7.2 Solving $\mathbf{P}_d$ via Barvinok's counting algorithm

Recall that the main idea in Barvinok's counting algorithm is to provide a compact description of the generating function $z \mapsto g(z) := \sum_{x \in \Omega(y) \cap \mathbb{Z}^n} z^x$ in the form

$$g(z) = \{\sum z^x \,|\, x \in \Omega(y) \cap \mathbb{Z}^n\} = \sum_{i \in I} \varepsilon_i \frac{z^{a_i}}{\prod_{k=1}^n (1 - z^{b_{ik}})}, \tag{7.2}$$

(with $\varepsilon_i \in \{-1, +1\}$) for some vectors $\{a_i, b_{ik}\}$ in $\mathbb{Z}^n$, and some index set I; see Section 4.2.

Solving $\mathbf{P}_d$ via Barvinok's counting algorithm + binary search

Barvinok's counting algorithm coupled with a standard dichotomy procedure yields an alternative to Lenstra's algorithm for integer programming, with no rounding procedure. Namely:

Binary search algorithm:
Input: $A \in \mathbb{Z}^{m \times n}$, $y \in \mathbb{Z}^m$, $c \in \mathbb{Z}^n$.
Output: $f_d(y, c) := \max\{c'x : Ax \leq y;\, x \in \mathbb{Z}^n\}$

1. Let $\Omega(y) := \{x \in \mathbb{R}^n : Ax \leq y\}$. Set

$$b := \max\{c'x : Ax \leq y;\, x \in \mathbb{R}^n\}, \quad a := \min\{c'x : Ax \leq y;\, x \in \mathbb{R}^n\}$$

   and so $a \leq f_d(y, c) \leq b$.
2. Run Barvinok's counting algorithm to compute $q := |\Omega(y) \cap \mathbb{Z}^n|$. If $q = 0$ there is no feasible integer solution and $f_d(y, c) = -\infty$. If $q > 0$ then go to step 3.
3. While $b > a$ do the following:
   - Set $u := \lceil (b - a)/2 \rceil$ and compute $q := |\Omega(y) \cap \mathbb{Z}^n \cap \{x : u \leq c'x \leq b\}|$
   - If $q > 0$ then $u \leq f_d(y, c) \leq b$ and repeat with $(a, b) := (u, b)$
   - If $q = 0$ then $a \leq f_d(y, c) \leq u$ and repeat with $(a, b) := (a, u - 1)$
   - Return $f_d(y, c) = b$.

Remarkably, the above algorithm finds $f_d(y, c)$ in a time that is polynomial in the input size of the problem. However, in step 3 of this scheme, one must run Barvinok's algorithm to obtain the compact form (7.2) of the new generating function associated with the (new) polyhedron considered at each step in the dichotomy, which can also be quite time consuming. We next provide an alternative, which uses Barvinok's counting algorithm only once.

Solving $\mathbf{P}_d$ via the digging algorithm

We first provide an upper bound ρ^* on the optimal value of $\mathbf{P}_d$ by a simple inspection of Barvinok's formula (7.2); in addition, under some (easy to check) condition on the reward vector c, ρ^* is also the optimal value $f_d(y,c)$ of $\mathbf{P}_d$. If not, further exploration via the *digging algorithm* permits us to retrieve $f_d(y,c)$.

Consider the integer program $\mathbf{P}_d$ in (5.1) with associated convex polyhedron:

$$\Omega(y) := \{x \in \mathbb{R}^n | Ax = y \quad x \geq 0\}. \tag{7.3}$$

Recall that from the definition of $\widehat{f}_d(y,c)$ for problem $\mathbf{I}_d$, one has $\widehat{f}_d(y,c) = g(\mathrm{e}^c)$ with g as in (7.2), and with $\mathrm{e}^c = (\mathrm{e}^{c_1},\ldots,\mathrm{e}^{c_n}) \in \mathbb{R}^n$. Therefore, with $r \in \mathbb{N}$,

$$\widehat{f}_d(y,rc) = g(\mathrm{e}^{rc}) = \sum_{i \in I} \varepsilon_i \frac{(\mathrm{e}^r)^{c'a_i}}{\prod_{k=1}^n (1-(\mathrm{e}^r)^{c'b_{ik}})} \tag{7.4}$$

for some vectors $\{a_i, b_{ik}\}$ in $\mathbb{Z}^n$ and some index set I (assuming $c'b_{ik} \neq 0$ for all $i \in I$, $k = 1,\ldots,n$). See Section 4.2.

Next, making the change of variable $u := \mathrm{e}^r \in \mathbb{R}$, (7.4) gives

$$\widehat{f}_d(y,rc) = g(\mathrm{e}^{rc}) = \sum_{i \in I} \varepsilon_i \frac{u^{c'a_i}}{\prod_{k=1}^n (1-u^{c'b_{ik}})} = \sum_{i \in I} \varepsilon_i \frac{u^{c'a_i}}{Q_i(u)} =: h(u) \tag{7.5}$$

for some functions $\{Q_i\}$ of u. For every $i \in I$, let Γ_i be the set

$$\Gamma_i := \{k \in \{1,\ldots,n\} \,|\, c'b_{ik} > 0\}, \; \forall i \in I, \tag{7.6}$$

with cardinality $|\Gamma_i|$, and define the vector $v_i \in \mathbb{Z}^n$ by

$$v_i := a_i - \sum_{k \in \Gamma_i} b_{ik}, \quad i \in I. \tag{7.7}$$

If $\Gamma_i = \emptyset$ then we let $|\Gamma_i| = 0$ and $v_i := a_i$.

Theorem 7.1. *Assume that $\mathbf{P}_d$ in (5.1) has a feasible point $x \in \mathbb{Z}^n$ and a finite optimal value $f_d(y,c)$, and let g and $\widehat{f}_d$ be as in (7.4) with $c \in \mathbb{R}^n$ such that $c'b_{ik} \neq 0$ for all $i \in I$, $k = 1,\ldots,n$.*

(i) The optimal value $f_d(y,c)$ of the integer program $\mathbf{P}_d$ is given by

$$f_d(y,c) = \lim_{r \to \infty} \frac{1}{r} \ln \widehat{f}_d(y,rc) = \lim_{r \to \infty} \frac{1}{r} \ln g(\mathrm{e}^{rc}). \tag{7.8}$$

(ii) With v_i as in (7.7), let S^* be the set

$$S^* := \{i \in I \,|\, c'v_i = \rho^* := \max_{j \in I} c'v_j\}. \tag{7.9}$$

Then $\rho^* \geq f_d(y,c)$, and

$$\rho^* = f_d(y,c) \quad \text{if} \; \sum_{i \in S^*} \varepsilon_i (-1)^{|\Gamma_i|} \neq 0. \tag{7.10}$$

Proof. (i) With $z := \mathrm{e}^{rc}$ in the definition of $f(z)$, we have

$$g(\mathrm{e}^{rc}) = \widehat{f_d}(y, rc) = \left\{ \sum_x e^{rc'x} \,|\, x \in \Omega(y) \cap \mathbb{Z}^n \right\},$$

and so,

$$\begin{aligned} \mathrm{e}^{f_d(y,c)} = \mathrm{e}^{\max\{c'x \,|\, x \in \Omega(y) \cap \mathbb{Z}^n\}} &= \max\{\mathrm{e}^{c'x} \,|\, x \in \Omega(y) \cap \mathbb{Z}^n\} \\ &= \lim_{r\to\infty} \left(\sum_{x \in \Omega(y) \cap \mathbb{Z}^n} (\mathrm{e}^{c'x})^r \right)^{1/r} \\ &= \lim_{r\to\infty} g(\mathrm{e}^{rc})^{1/r} = \lim_{r\to\infty} \widehat{f_d}(y, rc)^{1/r}, \end{aligned}$$

and by continuity of the logarithm,

$$f_d(y,c) = \max\{c'x \,|\, x \in \Omega(y) \cap \mathbb{Z}^n\} = \lim_{r\to\infty} \frac{1}{r} \ln g(\mathrm{e}^{rc}) = \lim_{r\to\infty} \frac{1}{r} \ln \widehat{f_d}(y, rc).$$

(ii) From (a), one may hope to obtain $f_d(y,c)$ by just considering the leading terms (as $u\to\infty$) of the functions $u^{c'a_i}/Q_i(u)$ in (7.5). If the sum in (7.5) of the leading terms (with same power of u) does not vanish, then one obtains $f_d(y,c)$ by a simple limit argument as $u\to\infty$. From (7.4)–(7.5) it follows that

$$\frac{u^{c'a_i}}{Q_i(u)} \approx \frac{u^{c'a_i}}{\alpha_i u^{\rho_i}} = \frac{u^{c'a_i - \rho_i}}{\alpha_i}, \quad \text{as } u\to\infty,$$

where $\alpha_i u^{\rho_i}$ is the leading term of the function $Q_i(u)$ as $u\to\infty$. Again, from the definition of Q_i, its leading term $\alpha_i u^{\rho_i}$ as $u\to\infty$ is obtained with

$$\rho_i = \begin{cases} \sum_{k\in\Gamma_i} c'b_{ik} & \text{if } \Gamma_i \neq \emptyset \\ 0 & \text{otherwise,} \end{cases}$$

and its coefficient α_i is 1 if $\rho_i = 0$ and $(-1)^{|\Gamma_i|}$ otherwise.

Remembering the convention that $\sum_{k\in\Gamma_i} c'b_{ik} = 0$ and $(-1)^{|\Gamma_i|} = 1$ if $\Gamma_i = \emptyset$, we obtain

$$\varepsilon_i \frac{u^{c'a_i}}{Q_i(u)} \approx \varepsilon_i (-1)^{|\Gamma_i|} u^{c'(a_i - \sum_{k\in\Gamma_i} b_{ik})} \quad \text{as } u\to\infty.$$

Therefore, with S^* and ρ^* as in (7.9), if $\sum_{i\in S^*} \varepsilon_i (-1)^{|\Gamma_i|} \neq 0$ then

$$h(u) \approx u^{\rho^*} \sum_{i\in S^*} \varepsilon_i (-1)^{|\Gamma_i|} \quad \text{as } u\to\infty,$$

so that $\lim_{u\to\infty} \frac{1}{r} \ln g(\mathrm{e}^{rc}) = \rho^*$. This and (7.8) yields $f_d(y,c) = \rho^*$, the desired result. From the above analysis it easily follows that if $\sum_{i\in S^*} \varepsilon_i (-1)^{|\Gamma_i|} = 0$ then ρ^* is only an upper bound on $f_d(y,c)$. □

Observe that the vectors $a_i, \{b_{ik}\}$ in Barvinok's formula depend only on the polyhedron $\Omega(y)$. Therefore, (7.10) in Theorem 7.1(b) provides a simple (and easy to check) necessary and sufficient condition on the vector $c \in \mathbb{R}^n$, to ensure that the optimal value $f_d(y,c)$ of $\mathbf{P}_d$ is equal to ρ^* in (7.9), obtained directly from Barvinok's formula.

The digging algorithm. The interest of Theorem 7.1 is that the value ρ^* is obtained by simple inspection of (7.5), which in turn is obtained in the time polynomial in the input size of the polyhedron $\Omega(y)$ when the dimension n is fixed.

When $\sum_{i\in S^*} \varepsilon_i(-1)^{|\Gamma_i|} \neq 0$ then it also yields the optimal value $f_d(y,c)$ of $\mathbf{P}_d$. On the other hand, if $\sum_{i\in S^*} \varepsilon_i(-1)^{|\Gamma_i|} = 0$, i.e., the sum of the leading terms of the functions $u^{c'a_i}/Q_i(u)$ (with same power of u) vanishes, then one needs to examine the "next" leading terms, which requires a further and nontrivial analysis of each function $u^{c'a_i}/Q_i(u)$. This strategy has been implemented in the *digging* algorithm of De Loera et al. [42].

Binary search versus the digging algorithm. The digging algorithm and the binary search algorithm are compared in De Loera et al. [42] on a sample of 15 hard knapsack problems with 5 to 10 variables, taken from Aardal and Lenstra [2]. None of them could be solved with the 6.6 version of the powerful CPLEX software package. For 13 of those 15 difficult knapsack problems, the digging algorithm provides the optimal value $f_d(y,c)$ much faster than the binary search algorithm, despite the former having exponential time complexity in the input size of $\Omega(y)$, whereas the latter has a polynomial time complexity (in fixed dimension). Notice that the test (7.10) was passed in seven problems of the list.

7.3 The link with Gomory relaxations

Recall from Section 6.2 that the *Gomory relaxation* of $\mathbf{P}_d$ is defined with respect to an optimal basis σ^* of the linear program $\mathbf{P}$ associated with $\mathbf{P}_d$. That is, if $A_{\sigma^*} \in \mathbb{Z}^{m\times m}$ denotes the submatrix of A associated with σ^*, and $\pi_{\sigma^*} \in \mathbb{R}^m$ denotes an optimal solution of the LP dual $\mathbf{P}^*$ of $\mathbf{P}$, then the Gomory relaxation is the integer program:

$$\mathrm{IP}_{\sigma^*}(y): \quad \begin{array}{l} \max\limits_x \sum_{j\notin\sigma^*} (c_j - \pi_{\sigma^*}A_j)x_j \\ \text{s.t. } A_{\sigma^*}x_{\sigma^*} + \sum_{j\notin\sigma^*} A_j x_j = y, \\ x_{\sigma^*} \in \mathbb{Z}^m,\, x_j \in \mathbb{N},\, j \notin \sigma^* \end{array} \tag{7.11}$$

with optimal value denoted $\max \mathrm{IP}_{\sigma^*}(y)$. That is, $\mathrm{IP}_{\sigma^*}(y)$ is obtained from $\mathbf{P}_d$ by *relaxing* the nonnegativity constraint on the vector $x_{\sigma^*} \in \mathbb{Z}^m$.

If $\mathrm{IP}_{\sigma^*}(y)$ has an optimal solution $x = (x_{\sigma^*}, \{x_j\}) \in \mathbb{Z}^m \times \mathbb{N}^{n-m}$ with $x_{\sigma^*} \geq 0$, then x is an optimal solution of $\mathbf{P}_d$, and $f_d(y,c) = y'\pi_{\sigma^*} + \max \mathrm{IP}_{\sigma^*}(y)$, i.e., the Gomory relaxation is *exact*. This eventually happens when y is sufficiently large. See Section 6.2.

Let $y \in \gamma$ (for some chamber γ) and let Δ be the set of feasible bases $\{\sigma\}$ of the linear program $\mathbf{P}$. Let $x(\sigma) \in \mathbb{R}^n_+$ be the corresponding vertex of $\Omega(y)$ in (7.3). For every $\sigma \in \Delta$, let $C_\sigma \subset \mathbb{R}^n$ be the set

$$C_\sigma := \{x \in \mathbb{R}^n \,|\, Ax = y, \quad x_j \geq 0, \quad \forall j \notin \sigma\},\ \sigma \in \Delta. \tag{7.12}$$

Recall that if $y \in \gamma$ then $\Omega(y)$ is a simple polyhedron and so C_σ is nothing less than the tangent cone of $\Omega(y)$ at the vertex $x(\sigma)$; see Section 4.2

With g as in (7.2), Brion's formula (4.5) applied to the polyhedron $\Omega(y)$ reads

$$g(z) = \sum_{\sigma \in \Delta} h(C_\sigma, z) \tag{7.13}$$

with

$$h(C_\sigma, z) := \{\sum z^x \,|\, x \in C_\sigma \cap \mathbb{Z}^n\},\ \sigma \in \Delta.$$

So, evaluating $h(C_\sigma, z)$ is summing up z^x over $C_\sigma \cap \mathbb{Z}^n$, the *feasible set* of the Gomory relaxation associated with the basis σ. In principle, for fixed $z = \mathrm{e}^c \in \mathbb{R}^n$, the Gomory relaxation is defined only for an *optimal* basis as it would be unbounded for non optimal bases. However, recall that (7.13) is a sum of *formal series* and for each $\sigma \in \Delta$ taken separately, $z \in \mathbb{C}^n$ is understood as making $h(C_\sigma, z)$ *finite*; again see Section 4.2.

Then, as the Gomory relaxation $\mathrm{IP}_{\sigma^*}(y)$ provides an upper bound on $f_d(y,c)$ (and exactly $f_d(y,c)$ when y is sufficiently large), one may apply Theorem 7.1 to the integer program $\mathrm{IP}_{\sigma^*}(y)$ in (7.11), instead of $\mathbf{P}_d$ in (5.1).

So, when the dimension n is fixed, Barvinok's counting algorithm produces in a time that is polynomial in the input size of C_{σ^*}, the equivalent compact form of $h(C_{\sigma^*}, z)$,

$$h(C_{\sigma^*}, z) = \sum_{i \in I_{\sigma^*}} \varepsilon_i \frac{z^{a_i}}{\prod_{k=1}^{n}(1 - z^{b_{ik}})}, \tag{7.14}$$

where the above summation is over the unimodular cones in Barvinok's decomposition of C_{σ^*} into unimodular cones. There is much less work to do because now, in Brion's formula (7.13), we have only considered the term $h(C_{\sigma^*}, z)$ relative to the optimal basis $\sigma^* \in \Delta$ of the linear program $\mathbf{P}$.

When the condition on c in Theorem 7.1(b) is satisfied, one obtains the optimal value of the Gomory relaxation $\mathrm{IP}_{\sigma^*}(y)$ (hence the optimal value $f_d(y,c)$ of $\mathbf{P}_d$ for sufficiently large y) in a time that is polynomial in the input size of $\Omega(y)$ when the dimension n is fixed.

Single cone digging algorithm. The *single cone digging* algorithm is a variant of the digging algorithm, where Barvinok's counting algorithm is only used to compute (7.14) instead of (7.2). Much faster than the original digging algorithm, it failed to provide the optimal value $f_d(y,c)$ in only two out of all the hard knapsack instances considered in [42].

7.4 Notes

Theorem 7.1 and most of the material in this chapter is from Lasserre [92]. Barvinok's original counting algorithm relied on Lenstra's polynomial time algorithm for integer programming [103]; see, e.g., the discussion in Barvinok and Pommersheim [15, p. 21]. Later, Dyer and Kannan [49] showed that in Barvinok's original algorithm, one may remove this dependence by using instead a short vector computation via the LLL algorithm [104]. This is the version implemented in the

LattE software package [40, 41] with free access at http://www.math.ucdavis.edu/~latte/ For more details on solving hard knapsack problems (taken from Aardal et al. [2]) and a comparison between the *digging* algorithm and Barvinok's binary search algorithm, which both use the LattE software, the interested reader is referred to De Loera et al. [42]. It is worth noticing that other approaches based on lattice basis reduction techniques described in Aardal and Lenstra [2] have been very successful on hard knapsack problems. See also Eisenbrand [53] for a general integer programming algorithm with improved polynomial time complexity in fixed dimension.

Finally, let us mention the related work of Hosten and Sturmfels [65] for computing

$$\delta := \sup_{y \in \Delta} f(y,c) - f_d(y,c),$$

the maximal gap between the optimal values of $\mathbf{P}$ and $\mathbf{P}_d$, as the right-hand side y of $Ax = y$ ranges over vectors $y \in \Delta \subset \mathbb{Z}^m$ for which $f_d(y,c)$ has a finite value. By combining nicely an algebraic characterization with the use of generating functions (and Barvinok's algorithm for getting their compact form efficiently), they prove that one can obtain δ in time polynomial in the input size, when the dimension n is fixed.

Chapter 8
A Discrete Farkas Lemma

8.1 Introduction

We pursue the comparison between linear and integer programming from a *duality* point of view, and in this chapter, we provide a discrete analogue of the celebrated Farkas lemma in linear algebra.

Let $A \in \mathbb{Z}^{m\times n}$, $y \in \mathbb{Z}^m$ and consider the problem of existence of a solution $x \in \mathbb{N}^n$ of the system of linear equations

$$Ax = y, \tag{8.1}$$

that is, the existence of a *nonnegative integral* solution of the linear system $Ax = y$. For $m = 1$ and $A \in \mathbb{N}^n$, one retrieves the (old) *Frobenius problem* in number theory ((8.1) is also called the (unbounded) *knapsack* equation) for which many results have been known for a long time (e.g., see Ehrhart [52], Laguerre [85], Netto [115], and Mitrinovic et al. [111, Chapter XIV.21]). For instance, the function $y \mapsto \widehat{f_d}(y,0)$ that counts the solutions $x \in \mathbb{N}^n$ of $a'x = y$ is a *quasipolynomial* of degree $n-1$ (that is, a polynomial of y with periodic coefficients) whose period is the least common multiple (l.c.m.) of the a_j. The so-called *Frobenius number* is the first integer $y_0 \in \mathbb{N}$ such that there always exists an integral solution whenever $y > y_0$. For more recent results, the interested reader is referred to Beck et al. [19] and the many references therein.

The celebrated *Farkas lemma* in linear algebra states that

$$\{x \in \mathbb{R}^n_+ \,|Ax = y\} \neq \emptyset \Leftrightarrow \left[u \in \mathbb{R}^m \text{ and } A'u \geq 0\right] \Rightarrow y'u \geq 0. \tag{8.2}$$

To the best of our knowledge, there is no *explicit* discrete analogue of (8.2). Indeed, the (test) Gomory and Chvátal functions in Blair and Jeroslow [23] (see also Schrijver [121, Corollary 23.4b], Ryan and Trotter [120]) are defined implicitly and recursively and do not provide a test directly in terms of the data A, y. In the same vein, the elegant *superadditive duality* theory developed after the pioneering work of Gomory [58] in the 1960s states that (8.1) has a nonnegative integral solution $x \in \mathbb{Z}^n$ if and only if $f(y) \geq 0$ for all superadditive functions $f: \mathbb{Z}^m \to \mathbb{R}$, with $f(0) = 0$, and such that $f(A_j) \geq 0$ for all $j = 1, \ldots, n$. In Chapter 10, we will interpret the results of this chapter and the next one, in the light of this superadditive duality.

In this chapter, we provide a *discrete* analogue of Farkas lemma for (8.1) to have a solution $x \in \mathbb{N}^n$. Namely, when A and y have nonnegative entries, that is, when $A \in \mathbb{N}^{m\times n}$, $y \in \mathbb{N}^m$, we

J.B. Lasserre, *Linear and Integer Programming vs Linear Integration and Counting*,
Springer Series in Operations Research and Financial Engineering, DOI 10.1007/978-0-387-09414-4_8,

prove that (8.1) has a solution $x \in \mathbb{N}^n$ *if and only if* the polynomial $z \mapsto z^y - 1$ $(:= z_1^{y_1} \cdots z_m^{y_m} - 1)$ in $\mathbb{R}[z_1, \ldots, z_m]$ can be written as

$$z^y - 1 = \sum_{j=1}^{n} Q_j(z)(z^{A_j} - 1) = \sum_{j=1}^{n} Q_j(z)(z_1^{A_{1j}} \cdots z_m^{A_{mj}} - 1) \tag{8.3}$$

for some polynomials $\{Q_j\}$ in $\mathbb{R}[z_1, \ldots, z_m]$ with *nonnegative* coefficients. In other words,

$$\{x \in \mathbb{N}^n | Ax = y\} \neq \emptyset \Leftrightarrow z^y - 1 = \sum_{j=1}^{n} Q_j(z)(z^{A_j} - 1), \tag{8.4}$$

for some polynomials $\{Q_j\}$ in $\mathbb{R}[z_1, \ldots, z_m]$ with *nonnegative* coefficients. (Of course, the sufficiency part of the equivalence in (8.4) is the hard part of the proof.)

Moreover, we also show that the degree of the Q_j in (8.3) is bounded so that checking existence of an integer solution $x \in \mathbb{N}^n$ to $Ax = y$ reduces to checking whether some linear system has a nonnegative real solution, a linear programming problem. This result is also extended to the general case $A \in \mathbb{Z}^{m \times n}$, $y \in \mathbb{Z}^m$, but now the Q_j in (8.4) are Laurent polynomials instead of standard polynomials.

We call (8.4) a *Farkas lemma* because as (8.2), it states a condition in terms of the *dual* variables z associated with the constraints $Ax = y$. In addition, let $z := e^{\lambda}$ and notice that the basic terms $y'\lambda$ and $A'\lambda$ in (8.2) also appear in (8.4) via z^y, which becomes $e^{y'\lambda}$ and via z^{A_j}, which becomes $e^{(A'\lambda)_j}$. In fact, as in the discrete case, the continuous Farkas lemma (8.2) can be restated in the spirit of (8.4). Indeed, $Ax = y$ has a nonnegative solution $x \in \mathbb{R}^n$ if and only if the polynomial $\lambda \mapsto y'\lambda$ can be written as

$$\lambda \mapsto y'\lambda = \sum_{j=1}^{n} Q_j(\lambda)(A_j'\lambda)$$

for some polynomials $\{Q_j\}$ in $\mathbb{R}[\lambda_1, \ldots, \lambda_m]$ with *nonnegative* coefficients. This point of view is developed in Section 8.2

To show (8.4), we use the function $\widehat{f_d}(y, 0)$ already used in Chapters 4 and 5. Indeed, existence of a solution $x \in \mathbb{N}^n$ to (8.1) is equivalent to showing that $\widehat{f_d}(y, 0) \geq 1$ and by a detailed analysis of the complex integral (4.12), we prove that (8.3) is a *necessary and sufficient* condition on y to have $\widehat{f_d}(y, 0) \geq 1$.

8.2 A discrete Farkas lemma

Before proceeding to the general case $A \in \mathbb{Z}^{m \times n}$, we first consider the case $A \in \mathbb{N}^{m \times n}$ where A has only nonnegative entries.

The case $A \in \mathbb{N}^{m \times n}$

In this section we assume that $A \in \mathbb{N}^{m \times n}$ and so $y \in \mathbb{N}^m$, because otherwise, the set $\{x \in \mathbb{N}^n \, | Ax = y\}$ is obviously empty.

Theorem 8.1. *Let $A \in \mathbb{N}^{m\times n}$, $y \in \mathbb{N}^m$. Then the following two statements (i) and (ii) are equivalent:*
(i) The linear system $Ax = y$ has an integral solution $x \in \mathbb{N}^n$.
(ii) The real-valued polynomial $z \mapsto z^y - 1 := z_1^{y_1} \cdots z_m^{y_m} - 1$ can be written as

$$z^y - 1 = \sum_{j=1}^{n} Q_j(z)(z^{A_j} - 1) \tag{8.5}$$

for some real-valued polynomials $Q_j \in \mathbb{R}[z_1, \ldots, z_m]$, $j = 1, \ldots, n$, all of them with nonnegative coefficients.

In addition, the degree of the Q_j in (8.5) is bounded by

$$y^* := \sum_{j=1}^{m} y_j - \min_k \sum_{j=1}^{m} A_{jk}. \tag{8.6}$$

Proof. (ii) $\Rightarrow$ (i). Assume that $z^y - 1$ can be written as in (8.5) for some polynomials $\{Q_j\}$ with nonnegative coefficients $\{Q_{j\alpha}\}$, that is,

$$Q_j(z) = \sum_{\alpha\in\mathbb{N}^m} Q_{j\alpha} z^\alpha = \sum_{\alpha\in\mathbb{N}^m} Q_{j\alpha} z_1^{\alpha_1} \cdots z_m^{\alpha_m} \tag{8.7}$$

for finitely many nonzero (and nonnegative) coefficients $\{Q_{j\alpha}\}$. By Theorem 4.1 the number $\widehat{f_d}(y, 0)$ of integral solutions $x \in \mathbb{N}^n$ of the linear system of equations $Ax = y$ is given by

$$\widehat{f_d}(y,0) = \frac{1}{(2\pi i)^m} \int_{|z_1|=\gamma_1} \cdots \int_{|z_m|=\gamma_m} \frac{z^{y-e_m}}{\prod_{j=1}^n (1 - z^{-A_k})} dz,$$

where $\gamma \in \mathbb{R}_+^m$ satisfies $A' \ln \gamma > 0$. Writing z^{y-e_m} as $z^{-e_m}(z^y - 1 + 1)$ we obtain

$$\widehat{f_d}(y,0) = B_1 + B_2,$$

with

$$B_1 = \frac{1}{(2\pi i)^m} \int_{|z_1|=\gamma_1} \cdots \int_{|z_m|=\gamma_m} \frac{z^{-e_m}}{\prod_{j=1}^n (1 - z^{-A_k})} dz,$$

and

$$B_2 := \frac{1}{(2\pi i)^m} \int_{|z_1|=\gamma_1} \cdots \int_{|z_m|=\gamma_m} \frac{z^{-e_m}(z^y - 1)}{\prod_{j=1}^n (1 - z^{-A_k})} dz$$

$$= \sum_{j=1}^{n} \frac{1}{(2\pi i)^m} \int_{|z_1|=\gamma_1} \cdots \int_{|z_m|=\gamma_m} \frac{z^{-e_m} Q_j(z)}{\prod_{k\neq j}(1-z^{-A_k})} dz$$

$$= \sum_{j=1}^{n} \sum_{\alpha\in\mathbb{N}^m} \frac{Q_{j\alpha}}{(2\pi i)^m} \int_{|z_1|=\gamma_1} \cdots \int_{|z_m|=\gamma_m} \frac{z^{A_j+\alpha-e_m}}{\prod_{k\neq j}(1-z^{-A_k})} dz.$$

From (4.12) in Theorem 4.1 (with $y := 0$) we recognize in B_1 the number of solutions $x \in \mathbb{N}^n$ of the linear system of equations $Ax = 0$, so that $B_1 = 1$. Next, again from (4.12) in Theorem 4.1 (now with $y := A_j + \alpha$), each term

$$C_{j\alpha} := \frac{Q_{j\alpha}}{(2\pi i)^m} \int_{|z_1|=\gamma_1} \cdots \int_{|z_m|=\gamma_m} \frac{z^{A_j+\alpha-e_m}}{\prod_{k\neq j}(1-z^{-A_k})} dz$$

is equal to

$$Q_{j\alpha} \times \text{the number of integral solutions } x \in \mathbb{N}^{n-1}$$

of the linear system of equations $\widehat{A}^{(j)}x = A_j + \alpha$, where $\widehat{A}^{(j)}$ is the matrix in $\mathbb{N}^{m\times(n-1)}$ obtained from A by deleting its jth column. As by hypothesis each $Q_{j\alpha}$ is nonnegative, it follows that

$$B_2 = \sum_{j=1}^{n} \sum_{\alpha\in\mathbb{N}^m} C_{j\alpha} \geq 0,$$

so that $\widehat{f_d}(y,0) = B_1 + B_2 \geq 1$. In other words, the linear system of equations $Ax = y$ has at least one solution $x \in \mathbb{N}^n$.

(i) $\Rightarrow$ (ii). Let $x \in \mathbb{N}^n$ be a solution of $Ax = y$, and write

$$z^y - 1 = z^{A_1x_1} - 1 + z^{A_1x_1}(z^{A_2x_2} - 1) + \cdots + z^{\sum_{j=1}^{n-1} A_jx_j}(z^{A_nx_n} - 1),$$

and

$$z^{A_jx_j} - 1 = (z^{A_j} - 1)\left[1 + z^{A_j} + \cdots + z^{A_j(x_j-1)}\right], \quad j = 1,\ldots,n,$$

if $x_j > 0$, to obtain (8.5) with $Q_1(z) = 1 + z^{A_1} + \cdots + z^{A_1(x_1-1)}$ if $x_1 > 0$, and

$$z \mapsto Q_j(z) := z^{\sum_{k=1}^{j-1} A_kx_k}\left[1 + z^{A_j} + \cdots + z^{A_j(x_j-1)}\right], \quad j = 2,3,\ldots,n.$$

We immediately see that each Q_j has all its coefficients nonnegative (and even in $\{0,1\}$).

Finally, the bound on the degree follows immediately from the proof of (i) $\Rightarrow$ (ii). □

Discussion

Denote by $s(u) := \binom{m+u}{u}$ the dimension of the vector space of polynomials in m variables, of degree at most u. In view of Theorem 8.1, given $y \in \mathbb{N}^m$ (and with y^* as in (8.6)), checking existence of a solution $x \in \mathbb{N}^n$ of the linear system of equations $Ax = y$ reduces to checking whether or not there exists a nonnegative *real* solution q to a system of linear equations $Mq = r$, with

- $n \times s(y^*)$ variables $q = \{q_{j\alpha}\}$, the nonnegative coefficients of the Q_j
- $s(y^* + \max_k \sum_{j=1}^m A_{jk})$ equations $(Mq)_i = r_i$, to identify the terms with same exponent in both sides of (8.5).

This in turn reduces to solving a linear program with $ns(y^*)$ variables and $s(y^* + \max_k \sum_j A_{jk})$ equality constraints. Observe that in view of (8.5), this linear program has a matrix of constraints M with only 0 and ± 1 coefficients. Moreover, M is very sparse as each of its rows contains at most $2n$ nonzero entries. In fact, as will be shown and used in the next chapter, M is totally unimodular.

From the proof of Theorem 8.1, one may even enforce the weights Q_j in (8.5) to be polynomials in $\mathbb{Z}[z_1, \ldots, z_m]$ (instead of $\mathbb{R}[z_1, \ldots, z_m]$) with nonnegative coefficients (and even with coefficients in $\{0, 1\}$). However, (a) above shows that the strength of Theorem 8.1 is precisely to allow $Q_j \in \mathbb{R}[z_1, \ldots, z_m]$ as it permits us to check feasibility by solving a (continuous) linear program. Enforcing $Q_j \in \mathbb{Z}[z_1, \ldots, z_m]$ would result in an *integer* program of size larger than that of the original problem.

Finally, if indeed $z^y - 1$ has the representation (8.4), then whenever $\lambda \in \mathbb{R}^m$ satisfies $A'\lambda \geq 0$, one has (letting $z := e^\lambda$)

$$e^{y'\lambda} - 1 = \sum_{j=1}^{n} Q_j(e^{\lambda_1}, \ldots, e^{\lambda_m}) \left[e^{(A'\lambda)_j} - 1 \right] \geq 0$$

(because all the Q_j have nonnegative coefficients), which implies $y'\lambda \geq 0$. Hence, we retrieve that $y'\lambda \geq 0$ whenever $A'\lambda \geq 0$, which is to be expected since of course, existence of integral solutions to (8.1) implies existence of real solutions.

The link with the Gröbner base approach

Theorem 8.1 reduces the issue of existence of a solution $x \in \mathbb{N}^n$ to a particular *ideal membership problem*, that is, $Ax = y$ has an integral solution $x \in \mathbb{N}^n$ if and only if the polynomial $z^y - 1$ belongs to the *binomial ideal* $I = \langle z^{A_j} - 1 \rangle_{j=1,\ldots,n} \subset \mathbb{R}[z_1, \ldots, z_m]$ with the additional condition that the weights Q_j all have *nonnegative* coefficients.

Interestingly, consider the ideal $J \subset \mathbb{R}[z_1, \ldots, z_m, w_1, \ldots, w_n]$ generated by the binomials $z^{A_j} - w_j$, $j = 1, \ldots, n$, and let $G := \{g_j\}$ be a Gröbner basis of J (for some given term ordering). Using the algebraic approach described in Adams and Loustaunau [4, §2.8], it is known that $Ax = y$ has a solution $x \in \mathbb{N}^n$ if and only if the monomial z^y is reduced (with respect to G) to some monomial w^α, in which case $\alpha \in \mathbb{N}^n$ is a feasible solution. That is,

$$z^y = w^\alpha + \sum_{g_j \in G} H_j g_j = w^\alpha + \sum_{k=1}^{n} W_k(z, w)(z^{A_j} - w_k), \tag{8.8}$$

for some polynomials $\{H_j, W_k\} \subset \mathbb{R}[z_1, \ldots, z_m, w_1, \ldots, w_n]$.

Observe that this is not a Farkas lemma as we do not know in advance $\alpha \in \mathbb{N}^n$ (we look for it!) to test whether $z^y - w^\alpha \in J$. One has to apply Buchberger's algorithm to (i) find a reduced Gröbner basis G of J and (ii) reduce z^y with respect to G and check whether the final result is a monomial

w^{α}. Moreover, note that the latter approach uses polynomials in n (primal) variables w and m (dual) variables z, in contrast with the (only) m dual variables z in Theorem 8.1.

In addition, nothing is said on the sign of the coefficients of the polynomials W_k in (8.8) obtained after the reduction with respect to G. However, observe that if in (8.8) one makes $w_j := 1$ for all $j = 1, \ldots, n$, one obtains

$$z^y - 1 = \sum_{k=1}^{n} W_k(z, 1, \ldots, 1)\,(z^{A_j} - 1),$$

as in (8.5); but we do not know whether the polynomial $W_k(z, 1, \ldots, 1) \in \mathbb{R}[z]$ has only nonnegative coefficients.

The 0-1 case

Suppose we now consider existence of solutions $x \in \{0,1\}^n$ instead of $x \in \mathbb{N}^n$ in the unbounded case. This is the same as solving (8.1) over $x \in \mathbb{N}^n$, with the additional constraints $x_i \le 1$ for all $i = 1, \ldots, n$. The latter constraints can in turn be replaced by the equality constraints $x_i + u_i = 1$ by adding n additional *slack* variables u_i, also constrained to be in $\mathbb{N}$.

Therefore, with $I \in \mathbb{N}^{n \times n}$ being the identity matrix, existence of solutions $x \in \{0,1\}^n$ to (8.1) reduces to existence of solutions $(x, u) \in \mathbb{N}^n \times \mathbb{N}^n$ for the system

$$\begin{bmatrix} A \mid 0 \\ - \quad - \\ I \mid I \end{bmatrix} \begin{bmatrix} x \\ - \\ u \end{bmatrix} = \begin{bmatrix} y \\ - \\ e_n \end{bmatrix} \tag{8.9}$$

with $\mathrm{e}_n = (1, \ldots, 1) \in \mathbb{N}^n$. As the matrix of the above linear system only has entries in $\mathbb{N}$, we are in a position to apply Theorem 8.1, that is,

Corollary 8.1. *Let $A \in \mathbb{N}^{m \times n}$, $y \in \mathbb{N}^m$. The following two statements (i) and (ii) are equivalent:*

(i) The linear system $Ax = y$ has an integral solution $x \in \{0,1\}^n$.
(ii) The polynomial $(z, w) \mapsto z^y w_1 \cdots w_n - 1 \in \mathbb{R}[z, w]$ can be written as

$$z^y w_1 \cdots w_n - 1 = \sum_{j=1}^{n} Q_j(z, w)\,(z^{A_j} w_j - 1) + \sum_{j=1}^{n} P_j(w)(w_j - 1) \tag{8.10}$$

for some polynomials $Q_j \in \mathbb{R}[z, w], P_j \in \mathbb{R}[w]$, all with nonnegative coefficients.

In addition, the degree of the polynomials Q_j in (8.10) is bounded by $y^* = y + n - \min_k \sum_{j=1}^{m} A_{jk}$, and that of P_j by n.

The special form of the weights Q_j, P_j in (8.10) is due to the fact that if $(x,u) \in \mathbb{N}^n \times \mathbb{N}^n$ is a solution of (8.9) then

$$\begin{aligned} z^y w_1 \cdots w_n - 1 &= \left[(z^{A_1} w_1)^{x_1} (z^{A_2} w_2)^{x_2} \cdots (z^{A_n} w_n)^{x_n} - 1\right] w_1^{u_1} \cdots w_n^{u_n} \\ &\quad + w_1^{u_1} \cdots w_n^{u_n} - 1 \\ &= w_1^{u_1} \cdots w_n^{u_n} \sum_{j=1}^n \tilde{Q}_j(z)(z^{A_j} w_j - 1) + w_1^{u_1} \cdots w_n^{u_n} - 1 \\ &= \sum_{j=1}^n Q_j(z,w)(z^{A_j} w_j - 1) + \sum_{j=1}^n P_j(w)(w_j - 1). \end{aligned}$$

The discrete Farkas lemma in the 0-1 case is considerably more complicated than for the unbounded case. Indeed, even if y is bounded by $\sum_j A_j$, one still has to search for n polynomials Q_j of degree at most $y + n - 1 - \min_k \sum_j A_{jk}$ in $2n$ variables, and n polynomials P_j in n variables of degree at most n. (Recall that a polynomial in n variables and of degree at most d has $\binom{n+d}{d}$ coefficients.)

Back to the continuous case

The classical (continuous) Farkas lemma in (8.2) can also be restated as an ideal membership problem, with special weights. Namely,

Proposition 8.1. *Let $A \in \mathbb{Z}^{m \times n}$. The following two statements are equivalent:*

(i) The linear system $Ax = y$ has a nonnegative solution $x \in \mathbb{R}^n$.
(ii) The (linear) polynomial $\lambda \mapsto y'\lambda \in \mathbb{R}[\lambda] \,(= \mathbb{R}[\lambda_1, \ldots, \lambda_m])$ can be written as

$$y'\lambda = \sum_{j=1}^n Q_j(\lambda)\,(A_j'\lambda), \tag{8.11}$$

for some polynomials $\{Q_j\}_{j=1}^n \subset \mathbb{R}[\lambda]$, all with nonnegative coefficients (denoted $Q_j \succeq 0$, for all $j = 1, \ldots, n$).

Proof. Suppose $Ax = y$ for some $x \in \mathbb{R}^n_+$. Then multiplying both sides with λ yields

$$y'\lambda = \langle Ax, \lambda \rangle = \langle x, A'\lambda \rangle = \sum_{j=1}^n x_j\,(A'\lambda)_j,$$

and so, (8.11) holds with $Q_j(\lambda) \equiv x_j \geq 0$, for all $j = 1, \ldots, n$. Conversely, assume that (8.11) holds for some polynomials $\{Q_j\} \subset \mathbb{R}[\lambda]$, all with nonnegative coefficients. Write

$$\lambda \mapsto Q(\lambda) = Q^0 + \sum_{0 \neq \alpha \in \mathbb{N}^m} Q^\alpha \lambda^\alpha,$$

with $\{Q^\alpha\} \subset \mathbb{R}^n_+$ for all $\alpha \in \mathbb{N}^n$. Substituting in (8.11) yields,

$$y'\lambda = \langle Q^0, A'\lambda \rangle + \sum_{0 \neq \alpha \in \mathbb{N}^n} \langle Q^\alpha \lambda^\alpha, A'\lambda \rangle, \; \forall \lambda \in \mathbb{R}^m.$$

Identifying terms of same degree in the above identity yields $y = AQ^0 = Ax$ with $x = Q^0 \geq 0 \in \mathbb{R}^n$. □

Clearly, Proposition 8.1 is the continuous analogue of Theorem 8.1 and (8.11) states that the linear polynomial $y'\lambda$ is in the ideal generated by the linear polynomials $\{A'_j\lambda\}$, with nonnegative weights $\{Q_j\} \subset \mathbb{R}[\lambda]$.

From the proof of Proposition 8.1, from any solution $\{Q_j\}_{j=1}^n \subset \mathbb{R}[\lambda]$ of (8.11) with $Q_j \succeq 0$ for all $j = 1, \ldots, n$, one obtains a solution of $Ax = y$ by taking $x_j := Q_j(0)$ for all $j = 1, \ldots, n$. In fact, Q_j must be a constant polynomial ($\equiv x_j \geq 0$) for all $j = 1, \ldots, n$. (Evaluate all partial derivatives at $\lambda = 0$ in both sides of (8.11) and use the fact that $Ax = 0$ with $x \geq 0$ yields $x = 0$.)

Finally, it is worth noticing that the continuous version (8.11) coincides with the *first-order* development of the discrete version (8.5). Indeed, expanding $z = e^\lambda$ in series in both sides of (8.5), and identifying linear terms, yields (8.11) for some polynomials $Q_j \succeq 0$, $j = 1, \ldots, n$, in $\mathbb{R}[\lambda]$.

A comparison between the continuous and discrete Farkas lemma is summarized in Table 8.1, where $\mathrm{conv}(\Omega(y) \cap \mathbb{Z}^n)$ denotes the *integer hull* of the convex polyhedron $\Omega(y)$, and $1_n \in \mathbb{R}^n$ denotes the vector of all ones.

Table 8.1 Comparing continuous and discrete Farkas lemma

$\Omega(y) = \{x \in \mathbb{R}^n_+ \mid Ax = y\}$	$\mathrm{conv}(\Omega(y) \cap \mathbb{Z}^n)$
$x \in \Omega(y)$ $\Leftrightarrow x = Q(0, \ldots, 0)$ with $Q \in \mathbb{R}[\lambda_1, \ldots, \lambda_m]^n$	$x \in \mathrm{conv}(\Omega(y) \cap \mathbb{Z}^n)$ $\Leftrightarrow x = Q(1, \ldots, 1)$ with $Q \in \mathbb{R}[e^{\lambda_1}, \ldots, e^{\lambda_m}]^n$
$y'\lambda = \langle Q, A'\lambda \rangle$, $Q \succeq 0$	$e^{y'\lambda} - 1 = \langle Q, e^{A'\lambda} - 1_n \rangle$, $Q \succeq 0$

Hence existence of a nonnegative solution $x \in \mathbb{R}^n_+$ (resp., $x \in \mathbb{Z}^n_+$) to a linear system $Ax = y$ is the same as existence of a nonnegative solution $0 \preceq Q \in \mathbb{R}[\lambda]$ (resp., $0 \preceq Q \in \mathbb{R}[z]$) to an abstract knapsack equation $y = \sum_{j=1}^n a_j Q_j$ in the polynomial ring $\mathbb{R}[\lambda]$ (resp., $\mathbb{R}[z]$) and where

- y is the polynomial $\lambda \mapsto y'\lambda$ (resp., $z \mapsto z^y - 1$)
- a_j is the polynomial $\lambda \mapsto (A'\lambda)_j$ (resp., $z \mapsto z^{A_j} - 1$), $j = 1, \ldots, n$.

The general case

In this section we consider the more general case where $A \in \mathbb{Z}^{m \times n}$ (so that A may have negative entries) and the convex polyhedron $\Omega(y) = \{x \in \mathbb{R}^n \mid Ax = y; x \geq 0\}$ is compact. The above arguments cannot be repeated because of the occurrence of negative powers. However, let $\alpha \in \mathbb{N}^n, \beta \in \mathbb{N}$ be such that

$$\widehat{A}_{jk} := A_{jk} + \alpha_k \geq 0, \quad \widehat{y}_j := y_j + \beta \geq 0, \quad k = 1,\ldots,n, j = 1,\ldots,m. \tag{8.12}$$

Moreover, as $\Omega(y)$ is compact,

$$\max_{x\in\mathbb{N}^n} \left\{ \sum_{j=1}^n \alpha_j x_j | Ax = y \right\} \leq \max_{x\in\mathbb{R}^n; x\geq 0} \left\{ \sum_{j=1}^n \alpha_j x_j | Ax = y \right\} =: \rho^*(\alpha) < \infty. \tag{8.13}$$

Observe that given $\alpha \in \mathbb{N}^n$, the scalar $\rho^*(\alpha)$ is easily calculated by solving a linear program. In (8.12) choose $\beta \geq \rho^*(\alpha)$, and let $\widehat{A} \in \mathbb{N}^{m\times n}$, $\widehat{y} \in \mathbb{N}^m$ be as in (8.12). Then existence of integral solutions $x \in \mathbb{N}^n$ for $Ax = y$ is equivalent to existence of integral solutions $(x,u) \in \mathbb{N}^n \times \mathbb{N}$ for the system of linear equations

$$\mathbf{Q}: \quad \begin{cases} \widehat{A}x + ue_m = \widehat{y} \\ \sum_{j=1}^n \alpha_j x_j + u = \beta. \end{cases} \tag{8.14}$$

Indeed, if $Ax = y$ with $x \in \mathbb{N}^n$ then

$$Ax + e_m \sum_{j=1}^n \alpha_j x_j - e_m \sum_{j=1}^n \alpha_j x_j = y + (\beta - \beta)e_m,$$

or equivalently,

$$\widehat{A}x + \left(\beta - \sum_{j=1}^n \alpha_j x_j \right) e_m = \widehat{y},$$

and thus, as $\beta \geq \rho^*(\alpha) \geq \sum_{j=1}^n \alpha_j x_j$ (see (8.13)), we see that (x,u) with $\beta - \sum_{j=1}^n \alpha_j x_j =: u \in \mathbb{N}$ is a solution of (8.14). Conversely, let $(x,u) \in \mathbb{N}^n \times \mathbb{N}$ be a solution of (8.14). Using the definitions of $\widehat{A}$ and $\widehat{y}$,

$$Ax + e_m \sum_{j=1}^n \alpha_j x_j + ue_m = y + \beta e_m, \quad \sum_{j=1}^n \alpha_j x_j + u = \beta,$$

and so, $Ax = y$. The system of linear equations (8.14) can be put in the form

$$B \begin{bmatrix} x \\ u \end{bmatrix} = \begin{bmatrix} \widehat{y} \\ \beta \end{bmatrix} \text{ with } B := \begin{bmatrix} \widehat{A} & | & e_m \\ - & & - \\ \alpha' & | & 1 \end{bmatrix}, \tag{8.15}$$

and as B has only entries in $\mathbb{N}$, we are back to the case analyzed in Section 8.2.

Theorem 8.2. *Let $A \in \mathbb{Z}^{m\times n}$, $y \in \mathbb{Z}^m$ and assume that $\Omega(y)$ is compact. Let $\widehat{A} \in \mathbb{N}^{m\times n}$, $\widehat{y} \in \mathbb{N}^m$, $\alpha \in \mathbb{N}^n$, and $\beta \in \mathbb{N}$ be as in (8.12) with $\beta \geq \rho^*(\alpha)$ (see (8.13)). Then the following two statements (i) and (ii) are equivalent:*

(i) The system of linear equations $Ax = y$ has an integral solution $x \in \mathbb{N}^n$.
(ii) The polynomial $(z,w) \mapsto z^y(zw)^\beta - 1 \in \mathbb{R}[z_1,\dots,z_m,w]$ can be written as

$$z^y(zw)^\beta - 1 = Q_0(z,w)(zw-1) + \sum_{j=1}^{n} Q_j(z,w)(z^{A_j}(zw)^{\alpha_j} - 1) \tag{8.16}$$

for some real-valued polynomials $\{Q_j\}_{j=0}^n$ in $\mathbb{R}[z_1,\dots,z_m,w]$, all with nonnegative coefficients.

In addition, the degree of the Q_j in (8.16) is bounded by

$$(m+1)\beta + \sum_{j=1}^{m} y_j - \min\left\{m+1, \min_{k=1,\dots,n}\left[(m+1)\alpha_k + \sum_{j=1}^{m} A_{jk}\right]\right\}.$$

Proof. Apply Theorem 8.1 to the equivalent form (8.15) of the system $\mathbf{Q}$ in (8.14), where B and $(\widehat{y},\beta)$ have only entries in $\mathbb{N}$, and use the definition (8.12) of $(\widehat{y},\beta)$ and $\widehat{A}$. □

We finally obtain the general form with Laurent polynomials instead of usual polynomials. Let $\widehat{y}$ and $\widehat{A}$ be as in (8.12) and let

$$t := \sup_{k=1,\dots,m} \frac{\widehat{y}_k}{\min_{j=1,\dots,n} \widehat{A}_{kj}}, \quad y^* := mt \sup_{k=1,\dots,m} \sum_{j=1}^{n} |A_{kj}|. \tag{8.17}$$

Corollary 8.2. *Let $A \in \mathbb{Z}^{m\times n}$, $y \in \mathbb{Z}^m$ be such that $\Omega(y)$ is compact. Then the two statements (i) and (ii) below are equivalent:*

(i) The system of linear equations $Ax = y$ has an integral solution $x \in \mathbb{N}^n$.
(ii) The Laurent polynomial $z \mapsto z^y - 1 \in \mathbb{R}[z,z^{-1}]$ can be written as

$$z^y - 1 = \sum_{j=1}^{n} Q_j(z)(z^{A_j} - 1) \tag{8.18}$$

for some Laurent polynomials $\{Q_j\}_{j=1}^n \subset \mathbb{R}[z,z^{-1}]$, all with nonnegative coefficients.

In addition, the degree of the Q_j in (8.18) is bounded by y^* in (8.17).

Proof. The proof mimics that of Theorem 8.1. The only difference is in the proof of (ii) → (i), where instead of (8.7) we now have

$$Q_j(z) = \sum_{\alpha\in\mathbb{Z}^m} Q_{j\alpha} z^\alpha = \sum_{\alpha\in\mathbb{Z}^m} Q_{j\alpha} z_1^{\alpha_1}\cdots z_m^{\alpha_m}$$

for finitely many nonzero coefficients $\{Q_{j\alpha}\}$. That is, the summation is over $\alpha \in \mathbb{Z}^m$ instead of $\alpha \in \mathbb{N}^m$.

Finally, concerning the degree bound y^*, observe that from (8.14), for any feasible solution $x \in \Omega(y) \cap \mathbb{Z}^n$, we have $x_j \le \widehat{y}_k/\widehat{A}_{kj}$ for all $k = 1,\dots,m$, and all $j = 1,\dots,n$. Therefore, $x_j \le t$ for all $j = 1,\dots,n$, with t as in (8.17). The bound y^* is easily obtained from the form of the Q_j in the part (i) $\Rightarrow$ (ii) of the proof of Theorem 8.1. □

8.3 A discrete theorem of the alternative

Theorem 8.1 states a discrete Farkas lemma in terms of some polynomial being or not being a member of a certain binomial ideal defined from the data A, y; see (8.5). Also, we have seen in Table 8.1 that the continuous standard Farkas lemma (8.2) can be rephrased in similar terms. In this section, we provide an equivalent formulation of Theorem 8.1 in the form of a theorem of the alternative, very much as is done for the continuous case (8.2).

As for Theorem 8.1, consider the case where A is nonnegative, that is, $A \in \mathbb{N}^{m\times n}$ and $y \in \mathbb{N}^m$, since the general case $A \in \mathbb{Z}^{m\times n}$, $y \in \mathbb{Z}^m$ reduces to the former case. In Theorem 8.1, checking whether (8.5) holds reduces to checking whether the polynomial $z \mapsto f(z) := z^y - 1$ is identical to the polynomial $z \mapsto h(z) := \sum_{j=1}^n Q_j(z)(z^{A_j} - 1)$ for some polynomials $(Q_j) \subset \mathbb{R}[z_1,\dots,z_m]$, all with nonnegative coefficients. There are several ways of expressing that f and h are identical.

For instance, one may evaluate both f and h at sufficiently many given points $\{z(i)\} \subset \mathbb{R}^n$ and state that $f(z(i)) = h(z(i))$ for all i. This yields a system of linear equations $\{Sq = s,\, q \ge 0\}$ that the nonnegative vector q of coefficients of $(Q_j) \subset \mathbb{R}[z]$ must satisfy.

Another possibility is to state that all partial derivatives of f and h agree when evaluated at some particular point $z = z^*$, i.e.,

$$\left.\frac{\partial^{|\alpha|} f}{\partial^{\alpha_1} z_1 \cdots \partial^{\alpha_m} z_m}\right|_{z=z^*} = \left.\frac{\partial^{|\alpha|} h}{\partial^{\alpha_1} z_1 \cdots \partial^{\alpha_m} z_m}\right|_{z=z^*}, \qquad \alpha \in \mathbb{N}^m. \tag{8.19}$$

This yields another system of linear equations that $q \ge 0$ must also satisfy. In fact, the system of linear equations $\{Mq = r,\, q \ge 0\}$ alluded to in the discussion that follows Theorem 8.1 is obtained from (8.19) with evaluation at $z^* = 0$. What if we do the same but now with evaluation at the point $z^* = 1$?

Let $b \in \mathbb{N}^m$ with $b > A_j$ for all $j = 1,\dots,n$, be fixed, arbitrary, and let $y \in \mathbb{N}^m$ with $y \le b$. Let $q = (q_j)$ where q_j is the nonnegative vector of coefficients of the polynomial $Q_j \in \mathbb{R}[z]$ in (8.5). From Theorem 8.1, for every $j = 1,\dots,n$, the total degree of Q_j is at most $(\sum_{k=1}^m y_k) - \min_j \sum_{k=1}^m A_{kj}$. In fact, more precisely, one may choose Q_j with monomials z^α such that $\alpha + A_j \le y$. Next, let

$$M^{(b)} q = s, \quad q \ge 0 \tag{8.20}$$

for some appropriate matrix $M^{(b)}$, be the linear system obtained from (8.19) with $y \le b$, $\alpha \ne 0$, and $z^* = 1$, and with the additional constraint $f(0) = h(0)$ (i.e., $1 = \sum_{j=1}^n Q_j(0)$); because 1 being a common root of f and h, the case $\alpha = 0$ in (8.19) yields no information. Hence, the right-hand side vector $s(y) = (s_\alpha(y))$ of (8.20) satisfies $s_0(y) = 1$, and

$$s_\alpha(y) = \left.\frac{\partial^{|\alpha|} z^y}{\partial^{\alpha_1} z_1 \cdots \partial^{\alpha_m} z_m}\right|_{z=1} = \prod_{\{j:\alpha_j>0\}} \prod_{k=1}^{\alpha_j} (y_j - k + 1) \tag{8.21}$$

for every $0 \neq \alpha \leq b$.

Notice that $y \mapsto s_\alpha(y)$ is a polynomial in y of degree at most $|\alpha|$. Next, let $C^{(b)}$ be the convex cone:

$$C^{(b)} := \{u : u^{\mathrm{T}} M^{(b)} \geq 0\}. \tag{8.22}$$

Theorem 8.3. *Let $A \in \mathbb{N}^{m\times n}$, $b \in \mathbb{N}^m$, and with $y \in \mathbb{N}^m$, $y \leq b$, let $s_\alpha \in \mathbb{R}[y]$ be as in (8.21) for every $\alpha \leq b$. Then the following two statements (i) and (ii) are equivalent:*

(i) The linear system $Ax = y$ has an integral solution $x \in \mathbb{N}^n$.
(ii) The linear system

$$\{u^{\mathrm{T}} M^{(b)} \geq 0; \quad u' s(y) < 0\} \tag{8.23}$$

has no real solution u.
Equivalently,

$$[\{x : Ax = y, y \in \mathbb{N}^m\} \neq \emptyset] \iff [u^{\mathrm{T}} M^{(b)} \geq 0 \Rightarrow u' s(y) \geq 0]. \tag{8.24}$$

Equivalently again, let $(u(k))_{k=1}^{r_b}$ be a set of generators of the convex cone $C^{(b)}$ defined in (8.22). Then

$$[\{x : Ax = y, y \in \mathbb{N}^m\} \neq \emptyset] \iff u(k)' s(y) \geq 0, \quad k = 1, \ldots, r_b. \tag{8.25}$$

Proof. By Theorem 8.1, the linear system $Ax = y$ has an integral solution $x \in \mathbb{N}^n$ if and only if (8.5) holds, which in turn is equivalent to stating that the linear system $M^{(b)} q = s$ has a nonnegative solution q. The standard Farkas lemma (8.2) applied to the latter yields (8.23) and the equivalent forms (8.24)–(8.25). □

Notice that we here obtain a theorem of the alternative in the spirit of the classical (8.2) for the continuous case. The difference between (8.2) and (8.25) is that in (8.25), $y \in \mathbb{N}^m$ must satisfy a set of *polynomial* inequalities instead of linear inequalities.

Example 8.1. Consider the knapsack equation $2x_1 + 5x_2 = 3$, $x_1, x_2 \in \mathbb{N}$, i.e., with $y = 3$ and $A = [2, 5] \in \mathbb{N}^{1\times 2}$. Let $b = 6$. Let $q = (q_1, q_2)$ with $q_1 = (q_{10}, \ldots, q_{14})$ and $q_2 = (q_{20}, q_{21})$. Let us consider the system of linear equations $M^{(b)} q = s$. One has $1 = q_{10} + q_{20}$, and with $0 < \alpha \leq b$ one obtains

$$y = 2\sum_{k=0}^{4} q_{1k} + 5(q_{20} + q_{21})$$

$$y(y-1) = 2\sum_{k=0}^{4} q_{1k} + 4\sum_{k=0}^{4} k q_{1k} + 20(q_{20} + q_{21}) + 10 q_{21}$$

$$y(y-1)(y-2) = 6\sum_{k=0}^{4} k(k-1)q_{1k} + 6\sum_{k=0}^{4} kq_{1k} + 60(q_{20}+q_{21}) + 60q_{21}$$

$$y(y-1)(y-2)(y-3) = 8\sum_{k=0}^{4} k(k-1)(k-2)q_{1k} + 12\sum_{k=0}^{4} k(k-1)q_{1k} + 120(q_{20}+q_{21}) + 240q_{21}$$

$$y(y-1)(y-2)(y-3)(y-4) = 240q_{14} + 120q_{13} + 480q_{14} + 600q_{21} + 120(q_{20}+q_{21})$$

$$y(y-1)(y-2)(y-3)(y-4)(y-5) = 720q_{14} + 720q_{21}.$$

With $y=1$, the second constraint yields $q_{jk}=0$ for all j,k, in contradiction with $1=q_{10}+q_{20}$. A dual certificate of this contradiction can be obtained with $u\in\mathbb{R}^7$ being such that $u_0=0$, $u_1=-1, u_2=-1/9$, and $u_k=0$, $k=3,\ldots,6$. Indeed, one has $M^{(b)}u\geq 0$ and at the point $y=1$, the polynomial $y\mapsto u_2y^2+(u_1-u_2)y$ is negative $(=u_1<0)$. By Theorem 8.3, $Ax=1$ has no integral solution $x\in\mathbb{N}^n$.

With $y=3$, the fourth constraint yields $q_{20}=q_{21}=0$ and $q_{1k}=0$, for $k\geq 2$. On the other hand, the second constraint yields $6=6q_{11}$ while the first constraint yields $6=2q_{10}+6q_{11}$, in contradiction with $1=q_{10}+q_{20}$.

On the other hand, with $y=4$, the fifth constraint yields $q_{13}=q_{14}=q_{20}=q_{21}=0$, and so, the fourth constraint yields $q_{12}=1$. Then the third constraint yields $q_{11}=0$, while $q_{10}=1$, and the second constraint yields $12=2q_{10}+10q_{12}$.

Finally, with $y=5$, the solution $q_{jk}=0$ (except $q_{20}=1$), and with $y=2$ the solution $q_{jk}=0$ (except $q_{10}=1$) are obviously feasible.

8.4 The knapsack equation

We now consider two particular cases of importance, the unbounded and 0-1 knapsack equations.

The unbounded knapsack equation. The (unbounded) knapsack (or Frobenius) equation is a particular case where $A'=a\in\mathbb{N}^n, y\in\mathbb{N}$, that is, $m=1$, and one considers the equation

$$a'x := \sum_{j=1}^{n} a_jx_j = y. \tag{8.26}$$

Therefore, as a direct consequence of Theorem 8.1, we obtain the next result.

Corollary 8.3. *Let $(a,y)\in\mathbb{N}^n\times\mathbb{N}$. The following two statements (i) and (ii) are equivalent:*

(i) The knapsack equation $a'x=y$ has a solution $x\in\mathbb{N}^n$.
(ii) The univariate polynomial $z\mapsto z^y-1$ in $\mathbb{R}[z]$ can be written as

$$z^y - 1 = \sum_{j=1}^{n} Q_j(z)(z^{a_j} - 1) \tag{8.27}$$

for some polynomials $Q_j \in \mathbb{R}[z]$ with nonnegative coefficients.

In addition, the degree of the Q_j in (8.27) is bounded by $y^* := y - \min_k a_k$.

From the discussion after Theorem 8.1, and in the present context of the knapsack equation, given $y \in \mathbb{N}$ one may test existence of a solution $x \in \mathbb{N}^n$ for (8.26) by solving a linear program with

- $\sum_k (y + 1 - a_k)$ variables (the unknown coefficients of the Q_j in (8.27));
- $(y+1)$ equality constraints, e.g., the linear constraints obtained from (8.19) in the previous section,

as illustrated in Example 8.1.

A particular case of importance. It is well known that if the a_j are *relatively prime* there are several explicit upper bounds for the *Frobenius number* y_0 such that (8.26) always has a solution $x \in \mathbb{N}^n$ whenever $y > y_0$. For instance as mentioned in Beck et al. [19], and with $a_1 < a_2 \cdots < a_n$, Erdös and Graham [55] provide the bound $2a_n \lfloor a_1/n \rfloor - a_1$, whereas Selmer [123] provides the bound $2a_{n-1} \lfloor a_n/n \rfloor - a_n$.

Therefore, if the a_j are relatively prime, to check whether (8.26) has a solution $x \in \mathbb{N}^n$ it suffices to consider only those y less than (for instance) Selmer's bound $2a_{n-1} \lfloor a_n/n \rfloor - a_n$. In this case, in view of Corollary 8.3, one has to solve a linear program with at most

- $2a_{n-1} \lfloor a_n/n \rfloor + 1 - a_n$ constraints;
- $n(2a_{n-1} \lfloor a_n/n \rfloor + 1 - a_n - a_1)$ variables.

The 0-1 knapsack equation. The 0-1 knapsack equation is the same as (8.26) except that now we search for solutions $x \in \{0,1\}^n$ instead of $x \in \mathbb{N}^n$ in the unbounded case. It is a particular example of the 0-1 case analyzed in (8.2), and so Corollary 8.1 takes the particular form:

Corollary 8.4. *Let $(a, y) \in \mathbb{N}^n \times \mathbb{N}$. The following two statements (i) and (ii) are equivalent:*

(i) The 0-1 knapsack equation $a'x = y$ has a solution $x \in \{0,1\}^n$.
(ii) The polynomial $(z, w) \mapsto z^y w_1 \cdots w_n - 1 \in \mathbb{R}[z, w]$ can be written as

$$z^y w_1 \cdots w_n - 1 = \sum_{j=1}^{n} Q_j(z, w)\,(z^{a_j} w_j - 1) + \sum_{j=1}^{n} P_j(w)(w_j - 1) \tag{8.28}$$

for some polynomials $Q_j \in \mathbb{R}[z, w]$ and $P_j \in \mathbb{R}[w]$, all with nonnegative coefficients.

In addition, the degree of the polynomials Q_j in (8.28) is bounded by $y^* = y + n - 1 - \min_k a_k$ and that of P_j by n.

8.5 Notes

Most of the material in this chapter is taken from Lasserre [90]. An alternative proof of Theorem 8.1 is provided in the next chapter and does not invoke the generating function approach. Let us mention the work of Ryan [118, 119] and Ryan and Trotter [120], who developed a general abstract Weyl–Minkowski duality. In particular, they show that given $A \in \mathbb{Z}^{m\times n}$, the integral monoid $S := \{y : y = Ax, x \in \mathbb{Z}^n_+\} \subset \mathbb{Z}^m$ generated by the columns of A is *finitely constrained* by Chvátal functions, i.e., $S := \{y \in \mathbb{Z}^m \,:\, f_j(y) \geq 0,\ j = 1,\ldots,r\}$ for some r Chvátal functions $\{f_j\}$.

Chapter 9
The Integer Hull of a Convex Rational Polytope

9.1 Introduction

Let $A \in \mathbb{Z}^{m\times n}$, $y \in \mathbb{Z}^m$, $c \in \mathbb{R}^n$ and consider the integer program $\mathbf{P}_d$ in (5.1) where the convex polyhedron $\Omega(y) = \{x \in \mathbb{R}^n \,|\, Ax = y, x \geq 0\}$ is compact. If $\mathrm{P} := \mathrm{conv}(\Omega(y) \cap \mathbb{Z}^n)$ denotes the *integer hull* of $\Omega(y)$, then solving $\mathbf{P}_d$ is equivalent to solving the linear program $\max\{c'x \,|\, x \in \mathrm{P}\}$.

However in general, finding the integer hull of a convex polyhedron is a difficult problem. As mentioned in Wolsey [134, p. 15], and to the best of our knowledge, no *explicit* (or "simple") characterization (or description) has been provided so far. For instance, its characterization via superadditive functions (e.g., as in Wolsey [134]) is mainly of theoretical nature. In the general *cutting plane* methods originated by Gomory and Chvátal in the early 1960s and some of the *lift-and-project* methods described in, e.g., Laurent [101], one obtains P as the final iterate of a *finite* nested sequence $\mathrm{P}^0 \supseteq \mathrm{P}' \supseteq \mathrm{P}'' \supseteq \cdots \supseteq \mathrm{P}$ of polyhedra. However, in all those procedures, there is no explicit description directly in terms of the initial data A, y. On the other hand, for specific polytopes $\Omega(y)$, one is often able to provide some *strong valid inequalities* in explicit form, but very rarely all of them (as for the matching polytope of a graph). For more details the interested reader is referred to Cornuejols and Li [37], Jeroslow [71], Laurent [101], Nemhauser and Wolsey [113], Schrijver [121, §23], and Wolsey [134, §8,9], and the many references therein.

In this chapter we first show that the integer program $\mathbf{P}_d$ in (5.1) is equivalent to a linear program in the *explicit* form

$$\max_{q\in\mathbb{R}^s}\{\widehat{c}\,'q | M q = r, \quad q \geq 0\}. \tag{9.1}$$

By "*explicit*" we mean that the data $M, r, \widehat{c}$ of the linear program (9.1) are constructed *explicitly* and *easily* from the initial data A, y, c. In fact, *no* calculation is needed and M, r have all their entries in $\{0, \pm 1\}$. In addition M is very sparse and *totally unimodular*. There is a simple linear relation $x = E\,q$ linking x and q (for some appropriate matrix E), but q is not a *lifting* of x, as in the lift-and-project procedures described in Laurent [101]. It is more appropriate to say that q is a *disaggregation* of x, as will become clear in the sequel.

We next provide a *structural* result on the integer hull P of the convex rational polytope $\Omega(y)$, in the sense that we obtain an *explicit* algebraic characterization of the defining hyperplanes of P in

J.B. Lasserre, *Linear and Integer Programming vs Linear Integration and Counting*,
Springer Series in Operations Research and Financial Engineering, DOI 10.1007/978-0-387-09414-4_9,

terms of generators of a convex cone C, which is itself described directly from the initial data A (via the matrix M in (9.1)), and with *no* calculation.

Of course, in view of the potentially large size of M, in general one cannot expect to get all generators of C. However, this structural result on the characterization of P may be helpful in either deriving strong valid inequalities or validating some candidate inequalities, at least for some specific polytopes $\Omega(y)$.

9.2 The integer hull

Denote by $e_m \in \mathbb{R}^m$ the vector with all entries equal to 1. By Theorem 8.1, and with same notation as in Chapter 8, if $A \in \mathbb{N}^{m\times n}$, $y \in \mathbb{N}^m$, the linear system $Ax = y$ has a solution $x \in \mathbb{N}^n$ if and only if the binomial $z^y - 1$ can be written as

$$z^y - 1 = \sum_{j=1}^{n} Q_j(z)(z^{A_j} - 1) \tag{9.2}$$

for some polynomials $\{Q_j\} \subset \mathbb{R}[z_1,\dots,z_m]$, all with *nonnegative* coefficients. Let $\{q_{j\alpha}\}$ denote the vector of coefficients of the polynomial Q_j (in the usual basis of monomials) for all $j = 1,\dots,n$.

From the discussion that follows Theorem 8.1, checking existence of an integral solution $x \in \mathbb{N}^n$ for the linear system $Ax = y$ is equivalent to checking whether there exists a nonnegative real solution q for the associated linear system

$$Mq = r; \quad q \geq 0 \tag{9.3}$$

for some matrix $M \in \mathbb{Z}^{p\times ns}$ and vector $r \in \mathbb{Z}^p$, with

- $s := \sum_{j=1}^{n} s_j$ variables $\{q_{j\alpha}\}$, the nonnegative coefficients $q_j = \{q_{\alpha j}\} \in \mathbb{R}^{s_j}$ of the polynomials Q_j, $j = 1,\dots,n$.
- p linear constraints that state that the two polynomials in both sides of (9.2) are identical. For instance, the constraints (8.19) with $z^* = 0$, which state that their coefficients are equal.

From the proof of Theorem 8.1, each polynomial $Q_j \in \mathbb{R}[z]$ in (9.2) may be restricted to contain only those monomials z^α with $\alpha \leq y - A_j$, $j = 1,\dots,n$. Therefore, we only need to identify terms z^α of same power in both sides of (9.2), with $\alpha \leq y$. That makes $p = \operatorname{card} \prod_{j=1}^{n}\{0,\dots,y_j\} = \prod_j (1 + y_j)$ constraints.

In addition, in view of (8.5), the matrix of constraints $M \in \mathbb{Z}^{p\times s}$, which has only 0 and ± 1 coefficients, is *easily* deduced from A with *no* calculation (and is very sparse). The same is true for $r \in \mathbb{Z}^p$ which has only two nonzero entries (equal to -1 and 1). Next, with M, r as in (9.3), let

$$\Delta := \{q \in \mathbb{R}^s \quad | \quad Mq = r; \quad q \geq 0\} \tag{9.4}$$

be the convex polyhedron of feasible solutions $q \in \mathbb{R}^s$ of (9.3).

For each $j = 1,\dots,n$, define the row vector $e_{s_j} := (1,\dots,1) \in \mathbb{R}^{s_j}$ and let $E \in \mathbb{N}^{n\times s}$ be the block diagonal matrix, with diagonal block $\{e_{s_j}\}$, that is,

$$E := \begin{bmatrix} e_{s_1} & 0 & \cdots & 0 \\ 0 & e_{s_2} & 0 & \cdots \\ 0 & \ldots & 0 & e_{s_n} \end{bmatrix}. \tag{9.5}$$

Theorem 9.1. *Let $A \in \mathbb{N}^{m \times n}, y \in \mathbb{N}^m$, and let Δ be the convex polyhedron defined in (9.4).*

(i) If $q \in \Delta$ then $x := Eq \in \Omega(y)$. In particular, if $q \in \Delta \cap \mathbb{Z}^s$ then $x \in \Omega(y) \cap \mathbb{Z}^n$.
(ii) Let $x \in \Omega(y) \cap \mathbb{Z}^n$. Then $x = Eq$ for some $q \in \Delta \cap \mathbb{Z}^s$.
(iii) The matrix M is totally unimodular.
(iv) If q is a vertex of $\Delta \neq \emptyset$, then $x = Eq \in \Omega(y) \cap \mathbb{Z}^n$.

Proof. (i) With $q \in \Delta$, let $\{Q_j\}_{j=1}^n \subset \mathbb{R}[z_1, \ldots, z_m]$ be the set of polynomials (with vector of nonnegative coefficients q) that satisfy (9.2). Taking the derivative of both sides of (9.2) with respect to z_j, at the point $z = (1, \ldots, 1)$, yields

$$y_k = \sum_{j=1}^n Q_j(1, \ldots, 1) A_{kj} = \sum_{j=1}^n A_{kj} x_j, \; j = 1, \ldots, n,$$

with $x_j := Q_j(1, \ldots, 1)$ for all $j = 1, \ldots, n$. Next, use the facts that (a) all the Q_j have nonnegative coefficients $\{q_{j\alpha}\}$ and (b) $Q_j(1, \ldots, 1) = \sum_{\alpha \in \mathbb{N}^m} q_{j\alpha} = (Eq)_j$ for all $j = 1, \ldots, n$, to obtain $x := Eq \in \Omega(y)$. Moreover, if $q \in \Delta \cap \mathbb{Z}^s$ then obviously $x \in \Omega(y) \cap \mathbb{Z}^n$ because E has integral entries.

(ii) Let $x \in \Omega(y) \cap \mathbb{Z}^n$ so that $x \in \mathbb{N}^n$ and $Ax = y$; write

$$z^y - 1 = z^{A_1 x_1} - 1 + z^{A_1 x_1} (z^{A_2 x_2} - 1) + \cdots + z^{\sum_{j=1}^{n-1} A_j x_j} (z^{A_n x_n} - 1),$$

and write

$$z^{A_j x_j} - 1 = (z^{A_j} - 1) \left[1 + z^{A_j} + \cdots + z^{A_j (x_j - 1)} \right], \; j = 1, \ldots, n,$$

whenever $x_j > 0$, to obtain (9.2) with $Q_j \equiv 0$ if $x_j = 0$,

$$Q_1(z) = 1 + z^{A_1} + \cdots + z^{A_1(x_1 - 1)} \quad \text{if } x_1 > 0,$$

and

$$Q_j(z) := z^{\sum_{k=1}^{j-1} A_k x_k} \left[1 + z^{A_j} + \cdots + z^{A_j (x_j - 1)} \right], \quad j = 2, 3, \ldots, n$$

if $x_j > 0$. We immediately see that each Q_j has all its coefficients $\{q_{j\alpha}\}$ nonnegative (and even in $\{0, 1\}$). Moreover, $Q_j(1, \ldots, 1) = x_j$ for all $j = 1, \ldots, n$, or equivalently, $x = Eq$ with $q \in \Delta \cap \mathbb{Z}^s$.

(iii) That M is totally unimodular follows from the fact that M is a *network matrix*, that is, a matrix with $\{0, \pm 1\}$ entries and with *exactly two* nonzero entries 1 and -1 in each column (see Schrijver [121, p. 274]). Indeed, from the identity (9.2), and the definition of M, each row of M is associated with a monomial z^α, with $\alpha \leq y$. Thus, consider a particular column of M

associated with the variable $q_{k\alpha}$ (the coefficient of the monomial z^{α} of the polynomial Q_k in (9.2), and so with $\alpha + A_k \leq y$). From (9.2), the variable $q_{k\alpha}$ is only involved

- in row (or equation) α associated with the monomial z^{α} (with -1 coefficient) and
- in row (or, equation) $\alpha + A_k (\leq y)$ associated with the monomial $z^{\alpha+A_k}$ (with $+1$ coefficient).

(iv) The right-hand side r in the definition of Δ is integral. Therefore, as M is totally unimodular, each vertex of Δ is integral (whenever $\Delta \neq \emptyset$). Therefore, if q is a vertex of Δ, one has $q \in \Delta \cap \mathbb{Z}^s$ and from (i), $x := Eq \in \Omega(y) \cap \mathbb{Z}^n$. □

From Theorem 9.1(b) and its proof, one sees that q is a disaggregation of $x \in \mathbb{N}^n$. Indeed, if we write $q = (q_1, \ldots, q_n)$ then each q_j has exactly x_j nontrivial entries, all equal to 1. So q is *not* a *lifting* of x as in the lift-and-project procedures described in Laurent [101]. In the latter, x is part of the vector q in the augmented space and is obtained by *projection* of q, whereas in Theorem 9.1, q is rather a disaggregation of x.

An equivalent linear program

Consider the integer program $\mathbf{P}_d$ in (5.1), and for every $c \in \mathbb{R}^n$ let $\widehat{c} \in \mathbb{R}^s$ be defined as

$$\widehat{c}\,' = (\widehat{c_1}\,', \ldots, \widehat{c_n}\,') \text{ with } \widehat{c}_j = c_j(1, \ldots, 1)' \in \mathbb{R}^{s_j} \quad \forall j = 1, \ldots, n. \tag{9.6}$$

Equivalently, $\widehat{c}\,' = c'E$ with E as in (9.5). It also follows that $\widehat{c}\,'q = c'x$ whenever $x = Eq$. As a consequence of Theorem 9.1 we obtain

Corollary 9.1. *Let $A \in \mathbb{N}^{m \times n}$, $y \in \mathbb{N}^m$, $c \in \mathbb{R}^n$.*

(i) The integer program

$$\mathbf{P}_d : \quad \max_x \{ c'x \,|\, Ax = y, \quad x \in \mathbb{N}^n \} \tag{9.7}$$

has same optimal value $f_d(y,c)$ as the linear program

$$\mathbf{Q} : \quad \max_{q \in \mathbb{R}^s} \{ \widehat{c}\,'q \,|\, Mq = r, \quad q \geq 0 \} \tag{9.8}$$

(including the case $-\infty$).

(ii) In addition, let $q^* \in \mathbb{R}^s$ be a vertex of Δ in (9.4), optimal solution of the linear program $\mathbf{Q}$. Then $x^* := Eq^* \in \mathbb{N}^n$ and x^* is an optimal solution of the integer program $\mathbf{P}_d$.

Proof. Let $f_d(y,c)$ and $\max \mathbf{Q}$ denote the respective optimal values of $\mathbf{P}_d$ and $\mathbf{Q}$. First consider the case $-\infty$. Then $f_d(y,c) = -\infty$ only if $\Omega(y) \cap \mathbb{Z}^n = \emptyset$. But then $\Delta = \emptyset$ as well, which in turn implies $\max \mathbf{Q} = -\infty$. Indeed, by Theorem 8.1, if $\Omega(y) \cap \mathbb{Z}^n = \emptyset$, i.e., if $Ax = y$ has no solution $x \in \mathbb{N}^n$, then one cannot find polynomials $\{\mathbf{Q}_j\} \subset \mathbb{R}[z_1, \ldots, z_m]$ with nonnegative coefficients that satisfy (9.2). Therefore, from its definition, $\Delta \neq \emptyset$ would yield a contradiction.

Conversely, if $\Delta = \emptyset$ (so that $\max \mathbf{Q} = -\infty$) then by definition of Δ, one cannot find polynomials $\{\mathbf{Q}_j\} \subset \mathbb{R}[z_1,\dots,z_m]$ with nonnegative coefficients that satisfy (9.2). Therefore, by Theorem 8.1, $Ax = y$ has no solution $x \in \mathbb{N}^n$, which in turn implies $f_d(y,c) = -\infty$, i.e., $\Omega(y) \cap \mathbb{Z}^n = \emptyset$.

In the case when $f_d(y,c) > -\infty$, we necessarily have $f_d(y,c) < \infty$ because the convex polyhedron $\Omega(y)$ is compact. Next, consider a feasible solution $q \in \Delta$ of $\mathbf{Q}$. From Theorem 9.1(i) $x := \mathrm{E}q \in \Omega(y)$. Therefore, as x is bounded then so is Eq, which, in view of the definition (9.5) of E, also implies that q is bounded. Hence Δ is compact, which in turn implies that the optimal value of $\mathbf{Q}$ is finite and attained at some vertex q^* of Δ.

Now, let $x^* \in \mathbb{N}^n$ be an optimal solution of $\mathbf{P}_d$. By Theorem 9.1(ii) there exists some $q \in \Delta$ with $Eq = x^*$. From the definition (9.6) of $\widehat{c}$, $\widehat{c}\,'q = c'E\,q = c'x^*$, which implies $\max \mathbf{Q} \geq f_d(y,c)$. On the other hand, let $q^* \in \Delta$ be a vertex of Δ, optimal solution of $\mathbf{Q}$. By Theorem 9.1(iv), $x := E\,q^* \in \Omega(y) \cap \mathbb{Z}^n$, that is, $x \in \mathbb{N}^n$ is a feasible solution of $\mathbf{P}_d$. Again, from the definition (9.6) of $\widehat{c}$, $c'x = c'E\,q^* = \widehat{c}\,'q^*$, which, in view of $f_d(y,c) \leq \max \mathbf{Q}$, implies $f_d(y,c) = \max \mathbf{Q}$, and $x \in \mathbb{N}^n$ is an optimal solution of $\mathbf{P}_d$. This completes the proof of (i) and (ii). □

Remark 9.1. On the dimension of $M \in \mathbb{Z}^{p\times s}$. Let $a := \sum_{j=1}^m y_j$. From the definitions of p and s, one has $p \leq \binom{m+a}{m} = p(m)$, and $s \leq n\binom{m+a}{m} = np(m)$, where $m \mapsto p(m)$ is a polynomial of degree a. Moreover, all the entries of M, r are $0, \pm 1$. Therefore, for the class $\mathscr{M}(A,y)$ of problems $\mathbf{P}_d$ where $A \in \mathbb{N}^{m\times n}$ and y is bounded, then so is a, and one may solve $\mathbf{P}_d$ in a time that is polynomial in the problem size, because it suffices to solve the linear program $\mathbf{Q}$, which has less than $p(m)$ constraints and $np(m)$ variables. One may consider this result a dual counterpart of the known result which states that integer programs are solvable in time polynomial in the input size when the dimension n is fixed.

The integer hull

We are now interested in describing the integer hull P of $\Omega(y)$, i.e., the convex hull of $\Omega(y) \cap \mathbb{Z}^n$.

Theorem 9.2. *Let $A \in \mathbb{N}^{m\times n}, y \in \mathbb{N}^m$, and let $E \in \mathbb{N}^{n\times s}, M \in \mathbb{Z}^{p\times s}, r \in \mathbb{Z}^p$ be as in (9.5) and (9.3), respectively.*

Let $\{(u^k,v^k)\}_{k=1}^t \subset \mathbb{R}^{n\times p}$ be a (finite) set of generators of the convex cone $C \subset \mathbb{R}^{n\times p}$ defined by

$$C := \{\,(u,v) \in \mathbb{R}^{n\times p} \mid E'u + M'v \geq 0\,\}. \tag{9.9}$$

The integer hull P of $\Omega(y)$ is the convex polyhedron defined by

$$\langle u^k, x\rangle + \langle v^k, r\rangle \geq 0 \;\forall k = 1,\dots,t, \tag{9.10}$$

or, equivalently,

$$\mathrm{P} := \{x \in \mathbb{R}^n \mid Ux \geq u\}, \tag{9.11}$$

where the matrix $U \in \mathbb{R}^{t\times n}$ has row vectors $\{u^k\}$, and the vector $u \in \mathbb{R}^t$ has coordinates $u_k = \langle -v^k, r\rangle$, $k = 1,\dots,t$.

Proof. Given $x \in \mathbb{R}^n$, consider the following linear system:

$$\begin{cases} Eq = x \\ Mq = r \\ q \geq 0 \end{cases} \tag{9.12}$$

where M, E are defined in (9.3) and (9.5), respectively. Invoking the continuous Farkas lemma (8.2), the system (9.12) has a solution $q \in \mathbb{R}^s$ if and only if (9.10) holds.

Therefore, let $x \in \mathbb{R}^n$ satisfy $Ux \geq u$ with U, u as in (9.11). By the Farkas lemma, the system (9.12) has a solution $q \in \mathbb{R}^s$, that is, $Mq = r, q \geq 0$ and $x = Eq$. As $q \in \Delta$ and Δ is compact, q is a convex combination $\sum_k \gamma_k \widehat{q}^{\,k}$ of the vertices $\{\widehat{q}^{\,k}\}$ of Δ. By Theorem 9.1(iv), for each vertex $\widehat{q}^{\,k}$ of Δ, one has $\widehat{x}^{\,k} := E\widehat{q}^{\,k} \in \Omega(y) \cap \mathbb{Z}^n$. Therefore,

$$x = Eq = \sum_k \gamma_k E\widehat{q}^{\,k} = \sum_k \gamma_k \widehat{x}^{\,k}, \tag{9.13}$$

that is, x is a convex combination of points $\widehat{x}^{\,k} \in \Omega(y) \cap \mathbb{Z}^n$, i.e., $x \in \mathrm{P}$; hence $\{x \in \mathbb{R}^n \,|\, Ux \geq u\} \subseteq \mathrm{P}$.

Conversely, let $x \in \mathrm{P}$, i.e., $x \in \mathbb{R}^n$ is a convex combination $\sum_k \gamma_k \widehat{x}^{\,k}$ of points $\widehat{x}^{\,k} \in \Omega(y) \cap \mathbb{Z}^n$. By Theorem 9.1(ii), for each k, $\widehat{x}^{\,k} = Eq^k$ for some vector $q^k \in \Delta \cap \mathbb{Z}^s$. Therefore, as each $(\widehat{x}^{\,k}, q^k)$ satisfies (9.12), then so does their convex combination $(x, q) := \sum_k \gamma_k (\widehat{x}^{\,k}, q^k)$. By the Farkas lemma again, we must have $Ux \geq u$, and so, $\mathrm{P} \subseteq \{x \in \mathbb{R}^n \,|\, Ux \geq u\}$, which completes the proof. □

Observe that the convex cone C in (9.9) of Theorem 9.2 is defined *explicitly* in terms of the initial data A, and with *no* calculation. Indeed, the matrix M in (9.3) is easily obtained from A and E is explicitly given in (9.5). Thus, the interest of Theorem 9.2 is that we obtain an algebraic characterization (9.11) of P via the generators of a cone C simply related to A.

From the proof of Theorem 9.2, every element (u, v) of the cone C produces a valid inequality for P_1, and clearly, all strong valid inequalities can be obtained from any set of *generators* of C.

Next suppose that for some $a \in \mathbb{R}^n$, $w \in \mathbb{R}$, we want to test whether $a'x \geq w$ is a valid inequality. If there is some $v \in \mathbb{R}^p$ such that $\mathrm{M}'v \geq -E'a$ and $w \leq -v'r$, then indeed, $a'x \geq w$ is a valid inequality. In fact w can be improved to $\tilde{w}$ with

$$\tilde{w} := \max_v \{-v'r | M'v \geq -E'a\}.$$

The general case $\mathbf{A} \in \mathbb{Z}^{\mathbf{m} \times \mathbf{n}}$

In this section we consider the case where $A \in \mathbb{Z}^{m \times n}$, that is, A may have negative entries. We will assume that the convex polyhedron $\Omega(y) \subset \mathbb{R}^n$ is compact.

We proceed as in Section 8.2. So let $\widehat{A} \in \mathbb{N}^{m \times n}$, $\widehat{y} \in \mathbb{N}^m$ be as in (8.12) with $\beta \geq \rho^*(\alpha)$ and $\rho^*(\alpha)$ as in (8.13). The feasible solutions $x \in \mathbb{N}^n$ of $Ax = y$, i.e., the points of $\Omega(y) \cap \mathbb{Z}^n$, are in one-to-one correspondence with the solutions $(x, u) \in \widehat{\Omega}(y) \cap \mathbb{Z}^{n+1}$ where $\widehat{\Omega}(y) \subset \mathbb{R}^{n+1}$ is the convex polytope

$$\widehat{\Omega}(y) := \left\{ (x, u) \in \mathbb{R}^{n+1} \,|\, B \begin{bmatrix} x \\ u \end{bmatrix} = (\widehat{y}, \beta), \quad x, u \geq 0 \right\}, \tag{9.14}$$

with

$$B := \begin{bmatrix} \widehat{A} & | & e_m \\ - & & - \\ \alpha' & | & 1 \end{bmatrix}.$$

(See Section 8.2.) As $B \in \mathbb{N}^{(m+1)\times(n+1)}$, we are back to the case analyzed in Theorem 9.1. In particular, the integer program $\mathbf{P}_d$: $\max\{c'x\,|\,Ax = y,\, x \in \mathbb{N}^n\}$ is equivalent to the integer program

$$\widehat{\mathbf{P}}_d : \quad \max\left\{c'x|B\begin{bmatrix} x \\ u \end{bmatrix} = \begin{bmatrix} \widehat{y} \\ \beta \end{bmatrix}, \quad (x,u) \in \mathbb{N}^n \times \mathbb{N}\right\}. \tag{9.15}$$

Hence, Theorem 8.1, Theorem 9.1, Corollary 9.1, and Theorem 9.2 are still valid with $B \in \mathbb{N}^{(m+1)\times(n+1)}$ in lieu of $A \in \mathbb{N}^{m\times n}$, $(\widehat{y},\beta) \in \mathbb{N}^m \times \mathbb{N}$ in lieu of $y \in \mathbb{N}^m$, and $\widehat{\Omega}(y) \subset \mathbb{R}^{n+1}$ in lieu of $\Omega(y) \subset \mathbb{R}^n$.

So again, as in previous sections, the polytope $\widehat{\Delta}$ associated with $\widehat{\Omega}(y)$ is explicitly defined from the initial data A, because $\widehat{A}$ is simply defined from A and α, and so, the matrices $\widehat{M}$ and $\widehat{E}$ are easily obtained from A and α.

In turn, as the convex cone $\widehat{C}$ in Theorem 9.2 is also defined explicitly from $\widehat{A}$ via $\widehat{M}$, one also obtains a simple characterization of the integer hull $\widehat{\mathrm{P}}$ of $\widehat{\Omega}(y)$ via the generators of $\widehat{C}$.

If we are now back to the initial data A, y, then P, is easily obtained from $\widehat{\mathrm{P}}$. Indeed, by Theorem 9.2, let

$$\widehat{\mathrm{P}} = \{(x,u) \in \mathbb{R}^{n+1} | \langle w^k, x\rangle + \delta^k u \geq \rho^k, \quad k = 1,\ldots,t\},$$

for some family $\{(w^k,\delta^k)\}_{k=1}^t \subset \mathbb{R}^n \times \mathbb{R}$ of generators of $\widehat{C}$. Then from (9.14), $u = \beta - \alpha'x$ and so,

$$\mathrm{P} = \{x \in \mathbb{R}^n | \langle w^k - \delta^k\alpha, x\rangle \geq \rho^k - \beta\delta^k, \quad k = 1,\ldots,t\}.$$

0-1 programs. Making the extension to 0-1 programs

$$\max_x\{c'x|Ax = y, \quad x \in \{0,1\}^n\}$$

is straightforward: Consider the equivalent integer program

$$\max_{x,u}\{c'x|Ax = y, \quad x_j + u_j = 1, \quad \forall j = 1,\ldots,n, \quad (x,u) \in \mathbb{N}^n \times \mathbb{N}^n\},$$

which is an integer program in the form $\mathbf{P}_d$ in (5.1). However, the resulting linear equivalent program $\mathbf{Q}$ of Corollary 9.1 is now more complicated. For instance, if $A \in \mathbb{N}^{m\times n}$, then $q \in \mathbb{R}^{2s}$ and s is now the dimension of the vector space of polynomials in $2n$ variables and of degree at most $n + \sum_j y_j$.

9.3 Notes

Most of the material in this chapter is from Lasserre [91]. Theorem 9.2 provides an explicit algebraic characterization of the integer hull P of the convex polytope $\Omega(y) \subset \mathbb{R}^n$ via generators of a polyhedral convex cone described explicitly in terms of A. Of course, and as expected, this convex cone is in a space of large dimension (exponential in the problem input size). However, this structural result shows that all *strong valid inequalities* can be obtained in this manner. Therefore, this result may be helpful in deriving strong valid inequalities, or in validating some candidate inequalities, at least for some specific polytopes $\Omega(y)$.

As already mentioned, there is also an explicit abstract description of the convex hull P in terms of superadditive functions. Namely,

$$\mathrm{P} = \left\{ x \in \mathbb{R}^n \ \sum_{j=1}^{n} f(A_j) x_j \leq f(y), \quad \forall f \in F, x \geq 0 \right\},$$

where F is the set of all superadditive functions $f : \mathbb{Z}^m \to \mathbb{R}$, with $f(0) = 0$; see, e.g., Wolsey [134]. In addition, Wolsey has also proposed a *superadditive* equivalent linear program to $\mathbf{P}_d$. Namely, there exists a finite set $\{F^k\}_{k=1}^r$ of superadditive functions $F_k : \mathbb{Z}^m \to \mathbb{R}$ such that the following program

$$\max \left\{ c'x : Ax = y, \quad \sum_{j=1}^{n} F_k(A_j) x_j \leq F_k(y),\, k = 1, \ldots, r,\, x \geq 0 \right\}$$

has the following properties:

- It has a feasible solution if and only if $\mathbf{P}_d$ has a feasible solution.
- It is unbounded if and only if $\mathbf{P}_d$ is unbounded.
- It has an optimal extreme point solution that is integral, and any such solution is optimal for $\mathbf{P}$.

In the same vein, there exists a finite set $\{G_k\}_{k=1}^s$ of superadditive functions $G_k : \mathbb{Z}^m \to \mathbb{R}$ such that the convex polyhedron

$$\left\{ Ax = y, \quad \sum_{j=1}^{n} G_k(A_j) x_j \leq G_k(y), \quad x \geq 0 \right\},$$

called the *y-hull* in Wolsey [135], is the convex hull of optimal solutions of $\mathbf{P}_d$ for all y.

In the next chapter we will see that results of the present chapter can be interpreted in the superadditive duality developed in, e.g., Johnson [74], Wolsey [133, 134], and the many references therein.

Chapter 10
Duality and Superadditive Functions

10.1 Introduction

In this chapter, we still consider the integer program $\mathbf{P}_d$ in (5.1) and relate results obtained in Chapters 8 and 9 with the so-called *superadditive dual* of $\mathbf{P}_d$ introduced in the late 1960s. Namely, related to $\mathbf{P}_d$ is the optimization problem

$$\min_{f\in\Gamma} \{\, f(y) \mid f(A_j) \geq c_j, \quad j=1,\ldots,n\,\}, \tag{10.1}$$

where Γ is the set of (extented) real-valued functions f on $\mathbb{Z}^m$ that are *superadditive* and such that $f(0)=0$. The problem (10.1) is the so-called *superadditive dual* of $\mathbf{P}_d$; see, e.g., Wolsey [134] (who considers the case $\{\,Ax \leq y, x \geq 0\,\}$). The reason (10.1) is a dual of $\mathbf{P}_d$ is clear almost immediately.

First, one may easily check that the value function $f_d(\cdot,c) : \mathbb{Z}^m \to \mathbb{R} \cup \{-\infty\}$ associated with $\mathbf{P}_d$ is superadditive and satisfies $f_d(0,c)=0$ and $f_d(A_j,c) \geq c_j$ for all $j=1,\ldots,n$ (assuming that $x=0$ is the only nonnegative solution of $Ax=0$). Next, if $f \in \Gamma$ is a feasible solution of (10.1) and $x \in \mathbb{N}^n$ satisfies $Ax=y$, then, by superadditivity of f,

$$f(y) = f\left(\sum_{j=1}^n A_j x_j\right) \geq \sum_{j=1}^n f(A_j x_j) \geq \sum_{j=1}^n f(A_j)x_j \geq \sum_{j=1}^n c_j x_j = c'x.$$

In other words, every feasible pair of solutions (f,x) of (10.1) and $\mathbf{P}_d$ satisfies $c'x \geq f(y)$, that is, the *weak duality* property holds. In addition, strong duality follows from $f_d(y,c) = \max \mathbf{P}_d$, and so $f_d(y,c)$ is an optimal solution of (10.1).

Except for the recent algorithmic work of Klabjan [80, 81], the dual problem (10.1) has remained rather conceptual in nature. However, one still retrieves several concepts already available in standard LP duality (see [134, p. 175] for the case $\{\,Ax \leq y, x \geq 0\,\}$). More important, the fundamental and basic Gomory (fractional) *cuts* for integer programs have an interpretation in terms of superadditive functions f in (10.1). Therefore, besides its theoretical interest, any insight on the dual problem (10.1) is of potential interest as it could provide useful information for deriving efficient cuts in solving any procedure for $\mathbf{P}_d$. Moreover, problem (10.1) (which is nonlinear) can be

J.B. Lasserre, *Linear and Integer Programming vs Linear Integration and Counting*,
Springer Series in Operations Research and Financial Engineering, DOI 10.1007/978-0-387-09414-4_10,

transformed into an equivalent finite LP. For instance, when $A \in \mathbb{N}^{m\times n}$, $y \in \mathbb{N}^m$, introducing the variables $\{\pi(\alpha)\}$ with $\alpha \in \mathbb{N}^m$ and $\alpha \leq y$, (10.1) is equivalent to

$$\begin{cases} \min_\pi & \pi(y) \\ \text{s.t.} & \pi(\beta+\alpha)-\pi(\alpha) \geq \pi(\beta), \qquad 0 \leq \alpha+\beta \leq y \\ & \pi(A_j) \geq c_j, \qquad\qquad\qquad j=1,\ldots,n \\ & \pi(0)=0. \end{cases} \tag{10.2}$$

The first set of constraints simply states that π is a superadditive function, whereas the second set of constraints is just that of (10.1); see also Wolsey [134, §2] for the inequality case.

How is the abstract dual (10.1) related to our results of Chapters 8 and 9? In Chapter 9, we have obtained a linear programming problem $\mathbf{Q}$ in (9.8) *equivalent* to $\mathbf{P}_d$ and which also yields a simple characterization of the *integer hull* P of the convex polytope $\Omega(y)$; see Corollary 9.1 and Theorem 9.2. It turns out that from every feasible solution of the LP dual $\mathbf{Q}^*$ of $\mathbf{Q}$, one may easily construct a feasible solution π of (10.2). In fact, $\mathbf{Q}^*$ is a *simplified* version of (10.2) in which we only have the constraints $\pi(A_j+\alpha)-\pi(\alpha) \geq c_j$ for all $\alpha+A_j \leq y$, and all j. The latter can in turn be interpreted as a longest path problem in a finite graph. We do not need to look for superadditive functions π as in (10.2), but from a feasible solution π of $\mathbf{Q}^*$, one may easily build up a superadditive function f_π feasible for (10.2). In particular, it shows that in the superadditive dual (10.1) of $\mathbf{P}_d$, one may restricted to the class of superadditive functions π coming from the representation (8.5) of the binomial z^y-1 in Theorem 8.1.

10.2 Preliminaries

As in Chapter 9 we may restrict ourselves to the case where $A \in \mathbb{Z}^{m\times n}$ has nonnegative entries, i.e., $A \in \mathbb{N}^{m\times n}$, and transpose results to the general case by the same simple transformation already shown in Section 8.2.

The superadditive dual of P

Let $A \in \mathbb{N}^{m\times n}$, $y \in \mathbb{N}^m$, and let $\mathbf{P}_d$ be the integer program (5.1) with optimal value $f_d(y,c): \mathbb{Z}^m \to \mathbb{R} \cup \{-\infty\}$ (with $f_d(y,c) := -\infty$ whenever $\Omega(y) \cap \mathbb{Z}^n = \emptyset$).

It follows that $f_d(0,c)=0$ (as $A \in \mathbb{N}^{m\times n}$) and $f_d(\cdot,c)$ is superadditive. Indeed, if $f_d(y,c) > -\infty$ and $f_d(y',c) > -\infty$ then we immediately have $f_d(y+y',c) \geq f_d(y,c)+f_d(y',c)$. If $f_d(y,c) = -\infty$ or $f_d(y',c) = -\infty$ (or both), then $f_d(y+y',c) \geq -\infty = f_d(y,c)+f_d(y',c)$.

Next, consider the abstract dual problem (10.1) where Γ is the set of functions $f: \mathbb{N}^m \to \mathbb{R} \cup \{-\infty\}$ such that $f(0)=0$ and f is superadditive. Let $f \in \Gamma$ be an arbitrary feasible solution, and let $x \in \mathbb{N}^n$ be such that $Ax=y$. Then,

$$f(y) = f\left(\sum_{j=1}^n A_j x_j\right) \geq \sum_{j=1}^n f(A_j x_j) \geq \sum_{j=1}^n f(A_j) x_j \geq \sum_{j=1}^n c_j x_j = c'x$$

(where both inequalities follow from the superadditivity of f), and so, $f(y) \geq c'x$, that is, the *weak duality* property holds. Strong duality holds with $f_d(y,c) = c'x^*$ for any optimal solution $x^* \in \mathbb{N}^n$

of $\mathbf{P}_d$. Next, if $Ax = y$ has no solution $x \in \mathbb{N}^n$, then $f_d(y,c) = -\infty$, in which case we also have $f_d(y,c) = \max \mathbf{P}_d$.

As already mentioned, the dual problem (10.1) is rather conceptual, as the set Γ is very large and not practical. For instance, Γ contains the value function $y \mapsto f_d(y,c)$ which is a bit too much of a requirement if one wishes to solve $\mathbf{P}_d$ for a single value y of the right-hand side. To make a parallel with linear programming, one may also define the dual of the linear program $\mathbf{P}$ in (3.1) to be (10.1), replacing now Γ with the set Γ' of *concave* functions $f : \mathbb{R}^n \to \mathbb{R} \cup \{-\infty\}$. The (large) set Γ' would also contain the value function $y \mapsto f(y,c)$ associated with $\mathbf{P}$. However, by considering the subclass of linear functions of Γ', one obtains the usual (and tractable) LP dual $\mathbf{P}^*$ of $\mathbf{P}$.

Similarly, the maximal valid inequalities that define the *integer hull* P of the convex polyhedron $\Omega(y)$ (see Chapter 9) can also be obtained from *Chvátal–Gomory cuts* formed from linear combinations and rounding of the facet inequalities that describe $\Omega(y)$. They are obtained from a subclass of superadditive functions that satisfy the constraints of (10.1); see, e.g., Nemhauser and Wolsey [113]. In Section 10.3 we also provide a characterization of P in terms of a finite family of superadditive functions.

A class of superadditive functions

Let $\mathscr{D} \subset \mathbb{N}^m$ be a finite set such that

$$0 \in \mathscr{D} \quad \text{and} \quad \alpha \in \mathscr{D} \quad \Rightarrow \quad \beta \in \mathscr{D} \quad \forall \beta \leq \alpha, \tag{10.3}$$

and let Δ be the set of functions $\pi : \mathbb{N}^m \to \mathbb{R} \cup \{+\infty\}$ such that

$$\pi \in \Delta \quad \text{if } \pi(0) = 0 \quad \text{and} \quad \pi(\alpha) = +\infty \quad \text{only if } \alpha \notin \mathscr{D}. \tag{10.4}$$

Next, given $\pi \in \Delta$ let $f_\pi : \mathbb{N}^m \to \mathbb{R} \cup \{+\infty\}$ be defined as

$$x \mapsto f_\pi(x) := \begin{cases} \inf_{\alpha + x \in \mathscr{D}} \{\pi(\alpha + x) - \pi(\alpha)\} & \text{if } x \in \mathscr{D} \\ +\infty \text{ otherwise.} \end{cases} \tag{10.5}$$

Observe that $f_\pi \in \Delta$ whenever $\pi \in \Delta$, that is,

$$f_\pi(0) = 0 \quad \text{and} \quad f_\pi(x) = +\infty \quad \text{only if } x \notin \mathscr{D}.$$

Indeed, if $x \in \mathscr{D}$ then from (10.5) $f_\pi(x) \leq \pi(x) - \pi(0) < +\infty$.

Proposition 10.1. *For every $\pi \in \Delta$*

(i) $f_\pi \leq \pi$, and f_π is superadditive;
(ii) if $\pi \in \Delta$ is superadditive, then $\pi = f_\pi$.

Proof. (i) $f_\pi(0) = 0$ follows from the definition (10.5) of f_π. Let $\pi \in \Delta$ so that $\pi(0) = 0$. If $x \notin \mathscr{D}$ then $f_\pi(x) = \pi(x) = +\infty$. If $x \in \mathscr{D}$ then $f_\pi(x) = \inf_{\alpha+x\in\mathscr{D}}\{\pi(\alpha+x)-\pi(\alpha)\} \leq \pi(x)-\pi(0) = \pi(x)$. Next, let $x,y \in \mathbb{N}^m$ be fixed, arbitrary. First consider the case where $x+y \in \mathscr{D}$, so that $x,y \in \mathscr{D}$.

$$\begin{aligned}
f_\pi(x+y) &= \inf_{\alpha+x+y\in\mathscr{D}} \{\pi(\alpha+x+y)-\pi(\alpha)\} \\
&= \inf_{\alpha+x+y\in\mathscr{D}} \{[\pi(\alpha+x+y)-\pi(\alpha+x)]+[\pi(\alpha+x)-\pi(\alpha)]\} \\
&\geq \inf_{\alpha+x+y\in\mathscr{D}} \{\pi(\alpha+x+y)-\pi(\alpha+x)\} + \inf_{\alpha+x\in\mathscr{D}} \{\pi(\alpha+x)-\pi(\alpha)\} \\
&= f_\pi(y)+f_\pi(x),
\end{aligned}$$

where we have used that $\alpha+x+y \in \mathscr{D}$ implies $\alpha+x \in \mathscr{D}$ (see (10.3)).

If $x+y \notin \mathscr{D}$, then from (10.5), $f_\pi(x+y) = +\infty$ and in this case one also has $f_\pi(x+y) \geq f_\pi(x)+f_\pi(y)$.

(ii) From (i) we have $f_\pi \leq \pi$. On the other hand, if $\pi \in \Delta$ is superadditive,

$$\pi(\alpha+x)-\pi(\alpha) \geq \pi(x) \quad \forall \alpha \in \mathbb{N}^m.$$

Therefore, if $x \in \mathscr{D}$ then

$$f_\pi(x) = \inf_{\alpha+x\in\mathscr{D}} \{\pi(\alpha+x)-\pi(\alpha)\} \geq \pi(x),$$

whereas $f_\pi(x) = +\infty = \pi(x)$ if $x \notin \mathscr{D}$. Combining this with $f_\pi \leq \pi$ yields the desired conclusion. □

10.3 Duality and superadditivity

With the integer program $\mathbf{P}_d$ in (5.1), we assume that $A \in \mathbb{N}^{m\times n}, y \in \mathbb{N}^m$. The general case with $A \in \mathbb{Z}^{m\times n}$ (and when $\Omega(y) = \{x \in \mathbb{R}^n \,|\, Ax = y;\, x \geq 0\}$ is compact) follows by reducing to the case $A \in \mathbb{N}^{m\times n}$ after a slight modification of the initial problem (introducing an additional constraint and an additional variable); see Section 9.2.

The link with superadditivity

From Corollary 9.1, recall that the linear program $\mathbf{Q}$ in (9.8) is equivalent to $\mathbf{P}_d$. To relate $\mathbf{Q}$ with superadditive functions, we proceed as follows.

Let $\mathscr{D} \subset \mathbb{N}^m$ be the set

$$\mathscr{D} := \prod_{j=1}^{m} \{0,1,\ldots,y_j\} = \{\,\alpha \in \mathbb{N}^m \mid \alpha \leq y\}. \tag{10.6}$$

In view of the simple form of the matrix M defined in Section 9.2, the LP dual $\mathbf{Q}^*$ of $\mathbf{Q}$ in (9.8) is easy to state. Namely, it is the linear program

$$\mathbf{Q}^* : \begin{cases} \min_\gamma & \gamma(y) - \gamma(0) \\ \text{s.t.} & \gamma(\alpha + A_j) - \gamma(\alpha) \geq c_j,\ \alpha + A_j \in \mathscr{D},\ j = 1, \ldots, n. \end{cases} \tag{10.7}$$

Clearly, by the change of variable $\pi(\alpha) := \gamma(\alpha) - \gamma(0)$, $\alpha \in \mathscr{D}$, (10.7) is equivalent to the LP

$$\mathbf{P}_d^* : \begin{cases} \min_\pi & \pi(y) \\ \text{s.t.} & \pi(\alpha + A_j) - \pi(\alpha) \geq c_j, \alpha + A_j \in \mathscr{D}, j = 1, \ldots, n \\ & \pi(0) = 0. \end{cases} \tag{10.8}$$

with optimal value denoted $\min \mathbf{P}_d^*$. Now, extend π to $\mathbb{N}^m$ by $\pi(\alpha) = +\infty$ whenever $\alpha \notin \mathscr{D}$. Then, with Δ as in (10.4), the linear program $\mathbf{P}_d^*$ is equivalent to the optimization problem

$$\begin{cases} \rho_1 := \min_{\pi \in \Delta} \pi(y) \\ \text{s.t. } \pi(\alpha + A_j) - \pi(\alpha) \geq c_j, \alpha + A_j \in \mathscr{D},\ j = 1, \ldots, n, \end{cases} \tag{10.9}$$

that is, $\min \mathbf{P}_d^* = \rho_1$.

Theorem 10.1. *Let $A \in \mathbb{N}^{m \times n}$, $y \in \mathbb{N}^m$, $c \in \mathbb{R}^n$, and let $\mathscr{D}$ be as in (10.6). Let $\mathbf{P}_d^*$ be the linear program defined in (10.8). Consider the optimization problem*

$$\begin{cases} \rho_2 := \inf_{\pi \in \Delta} f_\pi(y) \\ \text{s.t. } f_\pi(A_j) \geq c_j, j = 1, \ldots, n. \end{cases} \tag{10.10}$$

where $f_\pi : \mathbb{N}^m \to \mathbb{R}$ is the superadditive function defined in (10.5) for every $\pi \in \Delta$. If $\mathbf{P}_d$ is solvable, i.e., if $f_d(y, c) > -\infty$, then

$$f_d(y, c) = \min \mathbf{P}_d^* = \rho_2 = f_{\pi^*}(y) \quad \textit{for some } \pi^* \in \Delta. \tag{10.11}$$

Proof. First, by Corollary 9.1, $\max \mathbf{Q} = f_d(y, c)$, and by LP duality, $\max \mathbf{Q} = \min \mathbf{Q}^* = \min \mathbf{P}_d^*$. Thus, if $f_d(y, c) > -\infty$,

$$f_d(y, c) = \max \mathbf{Q} = \min \mathbf{P}_d^* = \rho_1.$$

Next, one proves that $\rho_1 \geq \rho_2$. Indeed, let $\pi \in \Delta$ be an optimal solution of (10.9). From its definition (10.5), f_π is an admissible solution of (10.10) because for all $j = 1, \ldots, n$,

$$f_\pi(A_j) = \inf_{\alpha + A_j \in \mathscr{D}} \pi(\alpha + A_j) - \pi(\alpha) \geq c_j.$$

In addition, from Proposition 10.1, $\pi \geq f_\pi$ and so, $\rho_1 = \pi(y) \geq f_\pi(y) \geq \rho_2$. Now, let $x \in \mathbb{N}^n$ be a feasible solution for $\mathbf{P}_d$, i.e., $Ax = y$, and with $\pi \in \Delta$, let f_π be a feasible solution in (10.10). Then

$$\begin{aligned}
f_\pi(y) = f_\pi\left(\sum_{j=1}^n A_j x_j\right) &\geq \sum_{j=1}^n f_\pi(A_j x_j) \quad \text{(by superadditivity)}\\
&\geq \sum_{j=1}^n f_\pi(A_j) x_j \quad \text{(by superadditivity)}\\
&\geq \sum_{j=1}^n c_j x_j = c'x.
\end{aligned}$$

This proves that $\rho_2 \geq f_d(y,c) = \rho_1$, and so $\rho_2 = f_d(y,c)$. □

Thus, a dual problem of $\mathbf{P}_d$ may have either the superadditive formulation (10.10) or the equivalent LP formulation $\mathbf{P}_d^*$ in (10.8). We further discuss the interpretation of the linear program $\mathbf{P}_d^*$ in Section 10.3.

Characterizing the integer hull via superadditivity

We now obtain a characterization of the integer hull P of the convex polytope $\Omega(y)$ via superadditive functions.

Let M, r, and E be as defined in (9.2) and recall from Chapter 9 that the integer hull P of $\Omega(y)$ is defined by

$$\mathrm{P} = \{x \in \mathbb{R}^n \mid \quad x'u^k + r'v^k \geq 0, \quad (u^k, v^k) \in \mathbf{G}\}, \tag{10.12}$$

where $\mathbf{G} \subset \mathbb{R}^n \times \mathbb{R}^p$ is a (finite) set of generators of the convex polyhedral cone $C \subset \mathbb{R}^n \times \mathbb{R}^p$, defined by

$$C = \{(u,v) \in \mathbb{R}^n \times \mathbb{R}^p \mid \quad M'v + E'u \geq 0\} \tag{10.13}$$

(see Theorem 9.2). From the definition of p, a vector $v \in \mathbb{R}^p$ may be viewed as a function $\pi : \mathscr{D} \to \mathbb{R}$, with $\mathscr{D}$ as in (10.6), and can be extended to $\mathbb{N}^m$ by setting $\pi(x) = +\infty$ if $x \notin \mathscr{D}$. Therefore, with a vector $v \in \mathbb{R}^p$ ($\sim \pi$) obtained from a generator $(u,v) \in \mathbf{G}$, we may and will associate a superadditive function f_π as in (10.5).

Theorem 10.2. *Let $A \in \mathbb{N}^{m\times n}, y \in \mathbb{N}^m$, $\mathscr{D}$ be as in (10.6), and let M, r, and E be as in (9.2). Let $\mathbf{G} \subset \mathbb{R}^n \times \mathbb{R}^p$ be a set of generators of the convex cone defined in (10.13). Then the integer hull P of the convex polytope $\Omega(y)$ is defined by*

$$\mathrm{P} = \left\{ x \in \mathbb{R}^n \mid \quad \sum_{j=1}^n f_\pi(A_j) x_j \leq f_\pi(y), \quad (u,\pi) \in \mathbf{G} \right\} \tag{10.14}$$

where f_π is the superadditive function obtained from π in (10.5) and $u_j = -f_\pi(A_j)$ for all $j = 1, \ldots, n$.

Proof. In view of the definition of M, r, and E, C in (10.13) is the convex cone

$$\{(u,\pi) \in \mathbb{R}^n \times \mathbb{R}^p \mid \quad \pi(\alpha+A_j) - \pi(\alpha) + u_j \geq 0, \quad \alpha + A_j \in \mathscr{D},\, j=1,\ldots,n\},$$

where the vector $\pi \in \mathbb{R}^p$ is viewed as a function $\pi : \mathscr{D} \to \mathbb{R}$, with $\mathscr{D}$ as in (10.6). Therefore, every generator $(u,\pi) \in \mathbf{G}$ satisfies

$$-u_j = \min_{\alpha+A_j \in \mathscr{D}} \{\pi(\alpha+A_j) - \pi(\alpha)\}, \quad j=1,\ldots,n.$$

Extend $\pi : \mathbb{N}^m \to \mathbb{R} \cup \{+\infty\}$ by $\pi(x) = +\infty$ if $x \notin \mathscr{D}$. Then, equivalently

$$-u_j = \min_{\alpha \in \mathscr{D}} \{\pi(\alpha+A_j) - \pi(\alpha)\} = f_\pi(A_j), \quad j=1,\ldots,n,$$

where f_π is the superadditive function obtained from π in (10.5). Also recall that from the definitions of $\mathscr{D}$ and f_π, one has $\pi(y) - \pi(0) = f_\pi(y)$; indeed, since $y + \alpha \notin \mathscr{D}$ whenever $\alpha \neq 0$, $\pi(y+\alpha) - \pi(\alpha) = +\infty > \pi(y) - \pi(0)$. Therefore, as $\pi' r = \pi(y) - \pi(0) = f_\pi(y)$, and $u_j = -f_\pi(A_j)$ for all $j = 1,\ldots,n$, we finally obtain

$$\mathrm{P} = \left\{ x \in \mathbb{R}^n \mid \quad \sum_{j=1}^{n} f_\pi(A_j) x_j \leq f_\pi(y), \quad (u,\pi) \in \mathbf{G} \right\},$$

which is (10.14). □

So, as in Nemhauser and Wolsey [113], the integer hull P is characterized in (10.14) via a *finite* family of superadditive functions. However, the superadditive functions in (10.14) are *not* obtained from linear combinations and rounding of the hyperplanes that define $\Omega(y)$, but rather from the set of generators $\mathbf{G}$ of the convex cone (10.13).

Discussion

Observe that in contrast with the LP (10.2), the linear program $\mathbf{P}_d^*$ in (10.8) does *not* optimize over the superadditive functions $\pi \in \Delta$, with Δ as in (10.4). In general, the constraints in (10.8) do not define a superadditive function π. But the function f_π obtained from π in (10.5) is superadditive and satisfies the constraints of the abstract dual (10.1).

Finally, note that the linear program $\mathbf{P}_d^*$ is simpler than (10.2). This is because the constraints of (10.2) state that $\pi : \mathscr{D} \to \mathbb{R}$ is superadditive, with $\pi(0) = 0$ and $\pi(A_j) \geq c_j$ for all $j = 1,\ldots,n$. So both linear programs (10.2) and $\mathbf{P}_d^*$ have the same number of variables. On the other hand, with $s := \prod_{j=1}^m (y_j + 1)$, the LP (10.2) has $O(s^2)$ constraints in contrast with $O(ns)$ constraints for $\mathbf{P}_d^*$.

Finally, $\mathbf{P}_d^*$ has an obvious interpretation as a large order problem in a graph G whose set of nodes is simply $\mathscr{D}$; a pair $(v,v') \in \mathscr{D} \times \mathscr{D}$ defines an arc of G if and only if $v' = v + A_j$ for some $j = 1,\ldots,n$, in which case its associated length is c_j. The goal is then to find the longest path between the nodes 0 and y.

Example 10.1. Let $A := [2,3] \in \mathbb{N}^{1\times 2}$ and $y := 5$ so that $Ax = y$ has only one solution $x^* = (1,1)$. Let the cost vector be $c = [c_1, c_2]$, and $\mathscr{D} := \{0,1,\ldots,5\}$. The dual problem $\mathbf{P}_d^*$ reads (with $\pi(0) = 0$)

$$\begin{cases} \min_\pi \ \pi(5) & \\ \text{s.t. } \pi(2) \geq c_1, & \pi(3) \geq c_2 \\ \pi(3) - \pi(1) \geq c_1, & \pi(4) - \pi(1) \geq c_2 \\ \pi(5) - \pi(3) \geq c_1, & \pi(5) - \pi(2) \geq c_2 \\ \pi(4) - \pi(2) \geq c_1, & \end{cases}$$

with optimal value $f_d(5,c) = \pi(5) = c_1 + c_2$, and optimal solution

$$\pi(1) = c_2 - c_1,\ \pi(2) = c_1,\ \pi(3) = c_2,\ \pi(4) = \max\{2c_1, 2c_2 - c_1\}.$$

The superadditive function $f_\pi : \mathbb{N}^2 \to \mathbb{R}$ defined in (10.5) (with $\pi(x) = +\infty$ if $x > 5$) satisfies $f_\pi(5) = c_1 + c_2$. For $y = 1$ the system $Ax = y$ has no solution and the LP dual $\mathbf{P}_d^*$ reads $\min_\pi \{\pi(1) \,|\, \pi(0) = 0\} = -\infty$, as $\alpha + A_j \notin \mathscr{D}$ for every $\alpha \in \mathscr{D} = \{0,1\}$. This is consistent with $f_d(1,c) = -\infty$.

We now summarize the main results of this chapter in Table 10.1 which compares the continuous and discrete optimization problems $\mathbf{P}$ and $\mathbf{P}_d$, in the case of a nonnegative constraint matrix $A \in \mathbb{N}^{m\times n}$ (and so, with $\mathscr{D}$ as in (10.6)).

Table 10.1 Problems $\mathbf{P}, \mathbf{P}_d$ and their respective duals

Primal Linear Program P	**Primal Integer Program $\mathbf{P}_d$**
$\mathbf{P}: \begin{array}{l} \max\ c'x \\ \text{s.t. } Ax = y \\ \quad x \geq 0, \quad x \in \mathbb{R}^n \end{array}$	$\mathbf{P}_d: \begin{array}{l} \max\ c'x \\ \text{s.t. } Ax = y \\ \quad x \geq 0, \quad x \in \mathbb{Z}^n \end{array}$
dual $\mathbf{P}^*$	dual $\mathbf{P}_d^*$
$\mathbf{P}^*: \begin{array}{l} \min_{\lambda \in \mathbb{R}^m}\ y'\lambda \\ \text{s.t. } A'\lambda \geq c. \end{array}$	$\mathbf{P}_d^*: \begin{array}{l} \min_{\pi \in \Delta}\ \pi(y) \\ \text{s.t. } \pi(\alpha + A_j) - \pi(\alpha) \geq c_j, \\ \alpha + A_j \in \mathscr{D};\ j = 1,\ldots,n, \\ \pi(0) = 0. \end{array}$
polytope $\Omega(y)$	integer hull $\mathrm{P} = \mathrm{co}(\Omega(y) \cap \mathbb{Z}^n)$
$\Omega(y): \begin{array}{l} \sum_{j=1}^n A_j x_j = y \\ x \geq 0 \end{array}$	$\mathrm{P}: \displaystyle\sum_{j=1}^n f_\pi(A_j) x_j \leq f_\pi(y), (u,\pi) \in \mathbf{G}$

In Table 10.1, one may also use the superadditive formulation (10.10), instead of $\mathbf{P}_d^*$. However, we have chosen the LP formulation $\mathbf{P}_d^*$ in (10.8) to better highlight the difference with the LP dual $\mathbf{P}^*$ of $\mathbf{P}$. Notice that if π is taken to be linear in $\mathbf{P}_d^*$, i.e., $\pi(\alpha) = \pi'\alpha$ for all $\alpha \in \mathscr{D}$ (for some vector $\pi \in \mathbb{R}^m$), then one retrieves exactly the LP dual $\mathbf{P}^*$ of $\mathbf{P}$.

10.4 Notes

The superadditive dual (10.1) dates back to the pioneering algebraic approach of Gomory [58] and later was further investigated and extended to mixed integer programs by several researchers, including Araoz [8], Blair [22], Blair and Jeroslow [23], Burdet and Johnson [30], Jeroslow [69, 71, 70], Johnson [74, 73], Klabjan [80, 81], Ryan [118, 119], Ryan and Trotter [120], Shapiro [124, 125], Tind and Wolsey [129], and Wolsey [133, 134, 135]. The earliest works were concerned with subadditive functions on finite groups, coming from an analysis of the group problem introduced in Gomory [58] (a relaxation of the original integer problem). The subaditive dual in Jeroslow [69] follows from a shortest path representation of integer programs (see also Shapiro [124]). It is then defined with subadditive functions on $\mathbb{Q}^m$, following the same reasoning as in Section 10.1, or in Wolsey [134], using the properties of the value function.

If in the abstract dual (10.1), one replaces superadditivity by *additivity*, then one retrieves the usual LP dual of the LP associated with $\mathbf{P}_d$ by dropping the integrality constraints. However, it is worth noting that the true abstract dual analogue of (10.1) for linear programs is obtained when one replaces superadditivity with *concavity*. Indeed, the value function of the LP associated with $\mathbf{P}_d$ is concave (and even piecewise linear), not linear, and the same reasoning as in Section 10.1 applies. However, LP duality shows that (for a fixed right-hand side y) one may be restricted to linear functionals and obtain a much simpler dual. But of course, the (concave) value function (for all values of y) cannot be a solution of the LP dual.

Appendix
Legendre–Fenchel, Laplace, Cramer, and $\mathbb{Z}$ Transforms

We briefly introduce some notions that are used extensively in the book. For a vector $x \in \mathbb{R}^n$, denote by x' its transpose and write without difference either $c'x$ or $\langle c, x\rangle$ for the scalar product of x and c.

An extended real-valued function $f : \mathbb{R}^n \to \mathbb{R} \cup \{-\infty, +\infty\}$ is said to be *lower semicontinuous* (l.s.c.) at $x \in \mathbb{R}^n$ if

$$\liminf_{y \to x} f(y) \geq f(x).$$

Equivalently, the *epigraph* $\operatorname{epi} f$ of f, defined by

$$\operatorname{epi} f := \{(x, r) \in \mathbb{R}^n \times \mathbb{R} : f(x) \leq r\},$$

is a closed subset of $\mathbb{R}^{n+1}$.

Given an extended real-valued function f, there is a greatest l.s.c. function majorized by f, which is called the *lower semicontinuous hull* of f.

An extended real-valued function $f : \mathbb{R}^n \to \mathbb{R} \cup \{-\infty, +\infty\}$ is said to be *convex* if $\operatorname{epi} f$ is a convex set. When $f(x) \neq -\infty$ for all $x \in \mathbb{R}^n$ then f is convex if and only if

$$f(\alpha x + (1-\alpha)y) \leq \alpha f(x) + (1-\alpha) f(y), \quad \alpha \in [0,1], \quad x, y \in \mathbb{R}^n.$$

The (effective) *domain* $\operatorname{dom} f \subset \mathbb{R}^n$ of f is the set of points where f is not $+\infty$. A convex function is said to be *proper* if $\operatorname{dom} f \neq \emptyset$ and $f(x) > -\infty$ for all $x \in \mathbb{R}^n$.

The *closure* $\operatorname{cl} f$ of a proper convex function f is its lower semicontinuous hull, and a proper convex function f is said to be *closed* if $\operatorname{cl} f = f$.

Given an extended real-valued function f, let $\operatorname{cl}(\operatorname{conv} f)$ denote the greatest closed convex function majorized by f. It is the supremum of the collection of all affine functions on $\mathbb{R}^n$ majorized by f.

A.1 The Legendre–Fenchel transform

Let $f : \mathbb{R}^n \to \mathbb{R} \cup \{-\infty, +\infty\}$ be given, and define $f^* : \mathbb{R}^n \to \mathbb{R} \cup \{-\infty, +\infty\}$ to be

$$\lambda \mapsto f^*(\lambda) := \sup_x \langle \lambda, x \rangle - f(x), \qquad \lambda \in \mathbb{R}^n. \tag{A.1}$$

The function f^* is the Legendre–Fenchel *conjugate* $\mathscr{F}[f]$ of f; it is convex and l.s.c. Notice that since f^* describes the affine functions majorized by f, f^* is the same as $(\mathrm{cl}\,(\mathrm{conv}\, f))^*$.

Proposition A.1. *Let f be convex. The conjugate $f^* = \mathscr{F}[f]$ is a closed convex function, proper if and only if f is proper. Moreover, $(\mathrm{cl}\, f)^* = f^*$ and $f^{**} = f$.*

So the conjugacy $f \to f^*$ induces a one-to-one mapping between closed proper convex functions. In particular, it is worth noticing that the conjugate of the (convex) quadratic $f = \frac{1}{2}x'Qx$ (with $Q \succ 0$) is the quadratic $f^* = \frac{1}{2}\lambda' Q^{-1}\lambda$.

The Fenchel–Young inequality states that

$$f(x) + f^*(\lambda) \geq \langle x, \lambda \rangle, \quad \forall x, \lambda \in \mathbb{R}^n,$$

and Fenchel–Young equality holds at (x,λ) if and only if $\lambda \in \partial f(x)$ and $x \in \partial f^*(\lambda)$, where $\partial f(x)$ denotes the subgradient of f at x.

An important property of the Legendre–Fenchel transform is to transform *inf-convolutions* into sums, and conversely, sums into inf-convolutions. That is, denoting $f \square g$ the inf-convolution of two convex l.s.c. functions f, g as

$$x \mapsto f \square g(x) := \inf_y \{f(y) + g(x-y)\},$$

one has

$$\mathscr{F}[f \square g] = \mathscr{F}[f] + \mathscr{F}[g], \quad \mathscr{F}[f+g] = \mathscr{F}[f] \square \mathscr{F}[g].$$

Example A.1. Let $x \mapsto f(x) = \frac{1}{2}x'Qx$ for some real-valued symmetric positive definite matrix $Q \in \mathbb{R}^{n\times n}$. Then $f^*(\lambda) = \frac{1}{2}\lambda' Q^{-1}\lambda$.

Example A.2. Let $A \in \mathbb{R}^{m\times n}$, $c \in \mathbb{R}^n$, and consider the linear program

$$\mathbf{P}: \min_x \{ \langle c, x \rangle \,|\, Ax = y; x \geq 0 \}, \quad y \in \mathbb{R}^m. \tag{A.2}$$

Define $f(\cdot, c) : \mathbb{R}^m \to \mathbb{R} \cup \{-\infty, +\infty\}$ to be the *value function* associated with $\mathbf{P}$, i.e.,

$$y \mapsto f(y,c) := \min_x \{ \langle c, x \rangle \,|\, Ax = y, x \geq 0 \}, \qquad y \in \mathbb{R}^m, \tag{A.3}$$

with the usual convention $f(y,c) = +\infty$ if $\Omega(y)\,(:= \{x \in \mathbb{R}^n \,|\, Ax = y; x \geq 0\}) = \emptyset$.

It is straightforward to check that f is convex, and in fact, convex and piecewise linear. Its conjugate f^* reads

$$\begin{aligned} f^*(\lambda,c) &= \sup_{y\in\mathbb{R}^m} \langle\lambda,y\rangle - f(y,c) \\ &= \sup_{y\in\mathbb{R}^m} \{\langle\lambda,y\rangle + \sup_x \{-c'x\,|\,Ax=y,\,x\geq 0\}\} \\ &= \sup_{x\geq 0} \langle A'\lambda - c, x\rangle \\ &= \begin{cases} 0 & \text{if } A'\lambda - c \leq 0 \\ +\infty & \text{otherwise} \end{cases} \end{aligned}$$

Next, we obtain

$$f^{**}(y,c) = \sup_{\lambda\in\mathbb{R}^m} \langle\lambda,y\rangle - f^*(\lambda,c) = \sup_{y\in\mathbb{R}^m} \{\langle\lambda,y\rangle\,|\,A'\lambda\leq c\}.$$

One recognizes in $f^{**}(\cdot,c)$ the value function of the dual linear program $\mathbf{P}^*$, and so $f(\cdot,c) = f^{**}(\cdot,c)$ as in Proposition A.1. Next, observe that $f^*(\cdot,c) = -\ln \mathrm{I}_{\Omega^*(c)}$, where $\mathrm{I}_{\Omega^*(c)}$ is the indicator function of

$$\Omega^*(c) := \{\lambda\in\mathbb{R}^m\,|A'\lambda\leq c\}, \tag{A.4}$$

the feasible set of the dual linear program $\mathbf{P}^*$. Next, from its definition we also have $-f(y,c) = \mathscr{F}[-\ln\mathrm{I}_{\Omega(y)}](-c)$. Therefore, we also obtain

$$(-f(y,\cdot))^*(-\gamma) = \mathscr{F}[\mathscr{F}[-\ln\mathrm{I}_{\Omega(y)}]](\gamma) = -\ln\mathrm{I}_{\Omega(y)}(\gamma), \quad \gamma\in\mathbb{R}^n,$$

and so, for every $\lambda\in\mathbb{R}^m, \gamma\in\mathbb{R}^n$,

$$(f(\cdot,c))^*(\lambda) = -\ln \mathrm{I}_{\Omega^*(c)}(\lambda) \quad \text{and} \quad (-f(y,\cdot))^*(-\gamma) = -\ln\mathrm{I}_{\Omega(y)}(\gamma).$$

A.2 Laplace transform

Before introducing the Laplace transform, we first recall the celebrated Cauchy residue theorem, a basic and fundamental result in complex analysis.

Cauchy residue theorem

Let $G\subset\mathbb{C}$ be an open set and $f:\mathbb{C}\to\mathbb{C}$ a given function. Then f is differentiable at $a\in G$ if

$$\lim_{h\to 0}\frac{f(a+h)-f(a)}{h}$$

exists. If f is differentiable at each point of G then f is differentiable on G.

A function $f : G \to \mathbb{C}$ is *analytic* if is continuously differentiable. One must recall that differentiability in $\mathbb{C}$ is a much stronger requirement than in $\mathbb{R}$. This can be realized when one sees that an analytic function is infinitely differentiable and can be expanded in a power series at each point of its domain!

If $\gamma : [0,1] : \to \mathbb{C}$ is a closed rectifiable curve and $a \notin \gamma$, then

$$n(\gamma;a) := \int_\gamma (z-a)^{-1}\, dz \in \mathbb{N}$$

is called the *index* of γ with respect to a. If $G \subset \mathbb{C}$ is an open set and $n(\gamma, w) = 0$ for all $w \in \mathbb{C} \setminus G$, then γ is said to be *homologous to zero*, denoted $\gamma \approx 0$.

Let f have an isolated singularity at $z = a$, and let

$$f(z) = \sum_{n=-\infty}^{\infty} a_n (z-a)^n$$

be its Laurent series expansion about $z = a$. Then the *residue* of f at $z = a$, denoted $\mathrm{Res}\,(f;a)$, is just the coefficient a_{-1}. So $\mathrm{Res}\,(f,a) = a_{-1}$.

We now come to a fundamental result in complex analysis.

Theorem A.1 (Cauchy Residue Theorem). *Let f be analytic on G, except for the isolated singularities $a_1, \ldots, a_m \in G$. If γ is a closed rectifiable curve in G that does not pass through any of the points a_k, $k = 1, \ldots, m$, and if $\gamma \approx 0$ in G then*

$$\frac{1}{2\pi i} \int_\gamma f\, dz = \sum_{k=1}^{m} n(\gamma, a_k)\, \mathrm{Res}\,(f, a_k).$$

In particular, if G is a disc with boundary circle C, and all the singularities are inside the disc, then

$$\frac{1}{2\pi i} \int_C f\, dz = \sum_{k=1}^{m} \mathrm{Res}\,(f, a_k).$$

The Laplace transform

The Laplace transform is a powerful technique which plays an important role in probability theory and in many applications. In particular, it is used in signal processing for analyzing linear time-invariant dynamical systems and solving linear ordinary differential equations where one reduces the initial problem to that of solving algebraic equations. In such physical systems it is interpreted as a transformation from the *time domain* to the *frequency domain*.

We start with the one-dimensional case. Let $f : \mathbb{R} \to \mathbb{R}$ be given, and such that

$$f = 0 \quad \text{on } (-\infty, 0) \quad \text{and} \quad |f(x)| \le \mathrm{e}^{ax} \quad \forall x \ge x_0,$$

for some $a > 0$. Then its *one-sided* Laplace transform $\mathscr{L}_1[f] : \mathbb{C} \to \mathbb{C}$ is defined by

$$\lambda \mapsto \mathscr{L}_1[f](\lambda) := \int_0^\infty e^{-\lambda x} f(x)\,dx, \qquad \lambda \in \mathscr{D}, \tag{A.5}$$

where $\mathscr{D} \subset \mathbb{C}$ is the domain where the above integral is well-defined. The right-hand side of (A.5) is called the *Laplace integral* of f, and typically $\mathscr{D} = \{\lambda \in \mathbb{C} : \Re(\lambda) > s\}$ for some $s \in \mathbb{R}$.

The inverse Laplace transform $\mathscr{L}_1^{-1}[f]$ is given by

$$x \mapsto \mathscr{L}_1^{-1}[f](x) := \frac{1}{2\pi i} \int_{\gamma - i\infty}^{\gamma + i\infty} e^{\lambda x} f(\lambda)\,d\lambda, \tag{A.6}$$

where $f : \mathbb{C} \to \mathbb{C}$ is analytic in the right half-plane $\Re(\lambda) > \gamma$.

The above integral is called the *Bromwich* integral, and in many cases of interest this integral is also the same as

$$\mathscr{L}_1^{-1}[f](x) = \frac{1}{2\pi i} \int_C e^{\lambda x} f(\lambda)\,d\lambda, \qquad x \geq 0, \tag{A.7}$$

where C is a closed contour of integration in $\mathbb{C}$, such that all singularities of f lie inside the contour. Then by using Cauchy residue theorem (Theorem A.1),

$$\mathscr{L}_1^{-1}[f](x) = \text{ sum of residues of } e^{\lambda x} f(x) \text{ at the poles of } f. \tag{A.8}$$

If $f : [0, \infty) \to \mathbb{R}$ is of bounded variation in a neighborhood of $x_0 \geq 0$, and if its Laplace integral converges absolutely on $\Re(\lambda) = c$, then

$$\mathscr{L}_1^{-1}[\mathscr{L}_1[f]] = \begin{cases} 0 & \text{if } x < 0, \\ f(0_+)/2 & \text{if } x = 0, \\ (f(x_+) + f(x_-))/2 & \text{if } x > 0, \end{cases}$$

where $f(x_+) = \lim_{y \downarrow x} f(y)$, and $f(x_-) = \lim_{y \uparrow x} f(y)$. And so if f is continuous, $\mathscr{L}_1^{-1}[\mathscr{L}_1[f]] = f$.

For $f : \mathbb{R} \to \mathbb{R}$, its *two-sided* (or *bilateral*) Laplace transform is given by

$$\mathscr{L}[f](\lambda) := \int_{-\infty}^\infty e^{-\lambda x} f(x)\,dx, \quad \lambda \in \mathscr{D}, \tag{A.9}$$

with $\mathscr{D} \subset \mathbb{C}$ being the domain where (A.9) makes sense. Its inverse is also given by the complex integral (A.6). Introducing the Heavyside *step* function

$$x \mapsto u(x) := \begin{cases} 0, & x < 0, \\ 1/2, & x = 0, \\ 1, & x > 0, \end{cases}$$

one may recover $\mathscr{L}_1[f]$ from $\mathscr{L}[f]$ by $\mathscr{L}_1[f] = \mathscr{L}[uf]$. The converse is also true. Indeed, defining $\tilde{f}$ by $\tilde{f}(x) = f(-x)$, one has

$$\mathscr{L}[f](\lambda) = \mathscr{L}_1[f](\lambda) + \mathscr{L}_1[\tilde{f}](-\lambda).$$

As the Legendre–Fenchel transform maps quadratics into quadratics, the Laplace transform maps Gaussian densities into Gaussian. Indeed

$$\mathscr{L}\left[\frac{\mathrm{e}^{-x^2/2\sigma}}{\sqrt{2\pi\sigma}}\right] = \mathrm{e}^{\lambda^2\sigma/2}.$$

There is an obvious extension to the multivariate case. For a given function $f : \mathbb{R}^m \to \mathbb{R}$, its bilateral Laplace transform $\mathscr{L}[f] : \mathbb{C}^m \to \mathbb{C}$ reads

$$\lambda \mapsto \mathscr{L}[f](\lambda) := \int_{\mathbb{R}^m} \mathrm{e}^{\langle\lambda,x\rangle} f(x)\,dx, \quad \lambda \in \mathscr{D},$$

where $\mathscr{D} \subset \mathbb{C}^m$ is the set of $\lambda \in \mathbb{C}^m$ such that the above integral is well-defined. With $x \in \mathbb{R}^m$, to recover $f(x)$ from the knowledge of $h := \mathscr{L}[f]$, one uses the inverse Laplace transform, also given by

$$f(x) = \mathscr{L}^{-1}[h](x) := \frac{1}{(2\pi i)^m} \int_{\gamma_1 - i\infty}^{\gamma_1 + i\infty} \cdots \int_{\gamma_m - i\infty}^{\gamma_m + i\infty} \mathrm{e}^{\langle\lambda,x\rangle} h(\lambda)\,d\lambda_1 \cdots d\lambda_m,$$

where $\gamma = (\gamma_1, \ldots, \gamma_m) \in \mathbb{R}^m$, $\{\gamma_1 \pm \infty\} \times \cdots \times \{\gamma_m \pm \infty\} \subset \mathscr{D}$, and h is analytic on $\Re(\lambda) > \gamma$.

One important property of the Laplace transform is to map convolutions into products. That is, denoting the convolution of f and g by

$$x \mapsto f \star g\,(x) := \int_{\mathbb{R}^n} f(x-y)\,g(y)\,dy,$$

one has

$$\mathscr{L}[f \star g] = \mathscr{L}[f] + \mathscr{L}[g].$$

Finally, observe that

$$\mathrm{e}^{\mathscr{F}[-f](-\lambda)} = \sup_x \mathrm{e}^{-\langle\lambda,x\rangle + f(x)} = \int^{\oplus} \mathrm{e}^{-\langle\lambda,x\rangle}\,\mathrm{e}^{f(x)},$$

where $\int^{\oplus}$ is the "integral" operator in the $(\max,+)$-algebra (recall that the "sup" operator is the $\oplus$ (addition) in the $(\max,+)$-algebra, i.e., $\max[a,b] = a \oplus b$. Formally, one has

$$\exp \mathscr{F}[-f] = \mathscr{L}\,[\exp f] \quad \text{in } (\max,+).$$

In other words, the Legendre–Fenchel and Laplace transforms are indeed the same transform interpreted in the $(\max,+)$- and $(+,\times)$-algebras, respectively.

The Cramer transform

The Cramer transform makes the link between the Legendre–Fenchel and Laplace transforms and is a useful tool to transform concepts from the usual $(+,\times)$-algebra to the $(\max,+)$-algebra. We have just seen that the Laplace and Legendre–Fenchel transforms are formal analogues in the respective $(+,\times)$- and $(\max,+)$-algebras.

One may observe that the *logarithmic Laplace transform* $\ln \mathscr{L}[f]$ of any nonnegative function $f : \mathbb{R}^n \to [0,+\infty]$ is convex and l.s.c. Therefore, it is appropriate to define its Legendre–Fenchel transform.

Given a function $f : \mathbb{R}^n \to [0,+\infty) \cup \{+\infty\}$, its Cramer transform $\mathscr{C}[f]$ is given by

$$f \mapsto \mathscr{C}[f] := \mathscr{F}[\ln \mathscr{L}[f]]. \tag{A.10}$$

In other words, $\mathscr{C} = \mathscr{F}[\ln \mathscr{L}]$ is the Legendre–Fenchel transform applied to the logarithm of the Laplace transform. As the $\ln \mathscr{L}[f]$ is convex and l.s.c., one has $\mathscr{C}[f]^* = \ln \mathscr{L}[f]$. As one may expect, since the Legendre–Fenchel transform maps quadratics into quadratics, and the Laplace transform maps Gaussians into Gaussians, the one-dimensional Cramer transform maps Gaussians into corresponding quadratics, i.e.,

$$\mathscr{C}\left[\frac{1}{\sqrt{2\pi\sigma^2}} \mathrm{e}^{(x-m)^2/2\sigma^2}\right](z) = \frac{(z-m)^2}{2\sigma^2}.$$

Next, as the logarithmic Laplace transform maps convolutions into sums, i.e.,

$$\ln \mathscr{L}[f \star g] = \ln \mathscr{L}[f] + \mathscr{L}[g],$$

the Cramer transform maps convolutions into inf-convolutions, i.e.,

$$\mathscr{C}[f \star g] = \mathscr{C}[f] \,\square\, \mathscr{C}[g]. \tag{A.11}$$

A.3 The $\mathbb{Z}$-transform

Just as one important application of the Laplace transform is to analyze *continuous-time* linear dynamical systems, one main application of the Z-transform is to analyze *discrete-time* linear dynamical systems. It converts a discrete-time domain signal (a sequence of real numbers) into a complex *frequency domain* representation.

The (bilateral) $\mathbb{Z}$-transform $\mathscr{Z}[f]$ of an infinite sequence $f = \{f(n)\}_{-\infty}^{\infty}$ is defined by

$$z \mapsto \mathscr{Z}[f](z) := \sum_{n=-\infty}^{\infty} f(n) z^{-n}, \qquad z \in \mathscr{D}, \tag{A.12}$$

where $\mathscr{D} \subset \mathbb{C}$ is the set of complex $z \in \mathbb{C}$ such that the above series converges. As already mentioned, the bilateral $\mathbb{Z}$-transform $\mathscr{Z}[f]$ is the obvious *discrete* analogue of the bilateral Laplace transform $\mathscr{L}[f]$. Similarly, the unilateral $\mathbb{Z}$-transform is defined as in (A.12), but with $f(n) = 0$ for all $n < 0$ and is also the discrete analogue of the unilateral Laplace transform. Uniqueness of the $\mathbb{Z}$-transform makes sense only if the domain $\mathscr{D}$ is specified.

In (A.12), the series converges in a ring of the form $0 \leq r_1 < z < r_2 \leq \infty$ where the radii r_1 and r_2 depend on the behavior of f at $\pm\infty$. If $f(n) = 0$ for all $n < 0$ (resp., $n > 0$) then $r_2 = +\infty$ (resp., $r_1 = 0$).

If $z = r\mathrm{e}^{i\theta}$ then $\mathscr{Z}[f]$ evaluated at $r = 1$ is the *Fourier transform*

$$\sum_{n=-\infty}^{\infty} f(n)\,\mathrm{e}^{-in\theta}$$

of the sequence f. If $h := \mathscr{Z}[f]$ is given by an algebraic expression and its domain of analyticity $\mathscr{D} \subset \mathbb{C}$ is known, then its inverse $\mathbb{Z}$-transform $\mathscr{Z}^{-1}[h]$ is obtained by

$$f(n) = \mathscr{Z}^{-1}[h](n) = \int_C h(z) z^{n-1}\, dz, \qquad n \in \mathbb{Z}, \tag{A.13}$$

where C is a closed contour in $\mathscr{D}$ surrounding the origin. Indeed, by the Cauchy residue theorem,

$$\int_C z^k z^{n-1}\, dz = \begin{cases} 1 & \text{if } k = -n \\ 0 & \text{otherwise,} \end{cases}$$

and so, replacing h in (A.13) with its expansion (A.12) yields the result.

The inverse $\mathbb{Z}$-transform (A.13) is the obvious *discrete* analogue of the inverse Laplace transform (A.7).

There is a straightforward extension to the multivariate case. Let $f = \{f(\alpha)\}$ be an infinite sequence indexed with $\alpha \in \mathbb{Z}^m$. Then, its $\mathbb{Z}$-transform $\mathscr{Z}[f]$ is now given by

$$z \mapsto \mathscr{Z}[f](z) := \sum_{\alpha \in \mathbb{Z}^m} f(\alpha) z^{-\alpha}, \qquad z \in \mathscr{D}, \tag{A.14}$$

where z^α is the usual notation for

$$z^\alpha = z_1^{\alpha_1} \cdots z_m^{\alpha_m}, \qquad \alpha \in \mathbb{Z}^m,$$

and $\mathscr{D} \subset \mathbb{C}^m$ is the set of $z \in \mathbb{C}^m$ such that the series in (A.14) converges. To recover $f(\alpha)$ from the knowledge of $h := \mathscr{Z}[f]$, one uses the multivariate inverse transform $\mathscr{Z}^{-1}$ given by

$$f(\alpha) = \mathscr{Z}^{-1}[h](\alpha) = \frac{1}{(2\pi i)^m} \int_{C_1} \cdots \int_{C_m} h(z)\, z^{\alpha - e}\, dz_1 \cdots dz_m,$$

with $\alpha \in \mathbb{Z}^m$, $z^{\alpha-e} = z_1^{\alpha_1 - 1} \cdots z_m^{\alpha_m - 1}$, and where $C_1 \times \cdots \times C_m \subset \mathscr{D}$, and each C_j is a closed contour in $\mathscr{D}$ encircling the origin.

A.4 Notes

For more details and applications on Laplace, Legendre–Fenchel, and Cramer transforms, as well as the links between them, the interested reader is referred to Bacelli et al. [10], Brychkov et al. [28], Burgeth and Weickert [31], and Litvinov and Maslov [106], and many references therein.

References

1. Aardal, K., Weismantel, R., Wolsey, L.A. Non-standard approaches to integer programming. *Discr. Appl. Math.* **123**, 5–74 (2002)
2. Aardal, K., Lenstra, A.K. Hard equality constrained integer knapsacks. In: W.J. Cook, A.S. Schulz, (eds.), Integer Programming and Combinatorial Optimization, pp. 350–366. Springer-Verlag, Heidelberg (2002)
3. Aardal, K., Lenstra, A.K. Hard equality constrained integer knapsacks. *Math. Oper. Res.* **29**, 724–738 (2004)
4. Adams, W.W., Loustaunau, P. An Introduction to Gröbner Bases. American Mathematical Society, Providence, RI (1994)
5. Akian, M., Quadrat, J.-P., Viot, M. Duality between probability and optimization. In: J. Gunawardena (ed.) Idempotency, Cambridge University Press, Cambridge (1998).
6. Alekseevskaya, T.V., Gel'fand, I.M., Zelevinsky, A.V. An Arrangement of real hyperplanes and the partition function connected with it. *Soviet Math. Dokl.* **36**, 589–593 (1988)
7. Allgower, E.L., Schmidt, P.M. Computing volumes of polyhedra. *Math. Comp.* **46**, 171–174 (1986)
8. Araoz, J. Polyhedral Neopolarities. PhD thesis, Dept. of Computer Sciences and Applied Analysis, Univ. of Waterloo, Waterloo, Ontario (1973)
9. Aráoz, J., Evans, L., Gomory, R.E., Johnson, E.L. Cyclic group and knapsack facets. *Math. Program. Ser. B* **96**, 377–408 (2003)
10. Bacelli, F., Cohen, G., Olsder, G.J., Quadrat, J-P. Synchronization and Linearity. John Wiley & Sons, Chichester (1992)
11. Baldoni-Silva, W., Vergne, M. Residues formulae for volumes and Ehrhart polynomials of convex polytopes, arXiv:math.CO/0103097 v1 (2001)
12. Baldoni-Silva, W., de Loera, J.A., Vergne, M. Counting integer flows in networks. *Found. Comput. Math.* **4**, 277–314 (2004)
13. Barvinok, A.I. Computing the volume, counting integral points and exponentials sums. *Discr. Comp. Geom.* **10**, 123–141 (1993)
14. Barvinok, A.I. A polynomial time algorithm for counting integral points in polyhedra when the dimension is fixed. *Math. Oper. Res.* **19**, 769–779 (1994)
15. Barvinok, A.I., Pommersheim, J.E. An algorithmic theory of lattice points in polyhedra. In: L.J. Billera (ed.), New Perspectives in Algebraic Combinatorics, pp. 91–147. MSRI Publications **38** (1999)
16. Barvinok, A.I., Woods, K. Short rational generating functions for lattice point problems. J. *Amer. Math. Soc.* **16**, 957–979 (2003)
17. Beck, M. Counting lattice points by means of the residue theorem. *Ramanujan J.* **4**, 399–310 (2000)
18. Beck, M. Multidimensional Ehrhart reciprocity. J. Comb. Theory Ser. A **97**, 187–194 (2002)
19. Beck, M., Diaz, R., Robins, S. The Frobenius problem, rational polytopes, and Fourier–Dedekind sums. *J. Number Theory* **96**, 1–21 (2002)
20. Beck, M. The partial-fractions method for counting solutions to integral linear systems. *Discrete Comput. Geom.* **32**, 437–446 (2004)
21. Bertsimas, D., Weismantel, R. Optimization over Integers. Dynamic Ideas, Belmont, Massachusetts (2005)
22. Blair, C.E. Topics in Integer Programming. PhD dissertation, Carnegie Mellon University (1975)

23. Blair, C.E., Jeroslow, R.G. The value function of an integer program, only *Math. Prog.* **23**, 237–273 (1982)
24. Bollobás, B. Volume estimates and rapid mixing. In: Levy, Silvio (eds.), Flavors of Geometry, pp. 151–180. Cambridge University Press, Cambridge **31** (1997)
25. Brion, M. Points entiers dans les polyèdres convexes. *Ann. Ecol. Norm. Sup. (Sér. 4)* **21**, 653–663 (1988)
26. Brion, M., Vergne, M. Lattice points in simple polytopes. *J. Amer. Math. Soc.* **10**, 371–392 (1997)
27. Brion, M., Vergne, M. Residue formulae, vector partition functions and lattice points in rational polytopes. *J. Amer. Math. Soc.* **10**, 797–833 (1997).
28. Brychkov, Y.A., Glaeske, H.J., Prudnikov, A.P., Tuan, V.K. Multidimensional Integral Transformations. Gordon and Breach, Philadelphia (1992)
29. Büeler, B., Enge, A., Fukuda, K. Exact volume computation for polytopes: A practical study. In: G. Kalai, G.M. Ziegler, (eds.), Polytopes—Combinatorics and Computati-on. Birkhäuser Verlag, Basel (2000)
30. Burdet, C.-A., Johnson, E.L. A subadditive approach to solve linear integer programs. *Ann. Discr. Math.* **1**, 117–144 (1977)
31. Burgeth, B., Weickert, J. An explanation for the logarithmic connection between linear and morphological systems. In: L.D. Griffin, M. Lillholm, (eds.), Scale-Space 2003, pp. 325–339. Lectures Notes in Computer Sciences, Springer-Verlag, Heidelberg (2003)
32. Chen, B. Lattice points, Dedekind sums, and Ehrhart polynomials of lattice polyhedra. *Discrete Comput. Geom.* **28**, 175–199 (2002)
33. Cochet, C. Réduction des graphes de Goretsky—Kottwitz—MacPherson; nombres de Kostka et coefficients de Littlewood—Richardson. Thèse de Doctorat, Université Paris 7, Paris, France (2003)
34. Cohen, J., Hickey, T. Two algorithms for determining volumes of convex polyhedra. *J. ACM* **26**, 401–414 (1979)
35. Conti, P., Traverso, C. Buchberger algorithm and integer programming. In: F.F. Mattson, T. Mora, T.R.N. Rao, (eds.), Applied Algebra, Algebraic Algorithms and Error-Correcting Codes, pp. 130–139. *Lecture Notes* in Computer Science **539**. Springer-Verlag, Heidelberg (1991)
36. Conway, J.B. Functions of a Complex Variable I. 2nd ed., Springer, New York (1978)
37. Cornuejols, G., Li, Y. Elementary closures for integer programs. *Oper. Res. Lett. ?* **28** 1–8 (2001)
38. Cook, W., Rutherford, T., Scarf, H.E., Shallcross, D. An implementation of the generalized basis reduction algorithm for integer programming, *ORSA J. Comp.* **5**, 206–212 (1993)
39. den Hertog, D. Interior Point Approach to Linear, Quadratic and Convex Programming. Kluwer Academic Publishers, Dordrecht (1994)
40. De Loera, J.A., Hemmecke, R., Tauzer, J., Yoshida, R. Effective lattice point counting in rational convex polytopes. *J. Symb. Comp.* **38**, 1273–1302 (2004)
41. De Loera, J.A., Haws, D., Hemmecke, R., Huggins, P., Tauzer, J., Yoshida, R. A User's Guide for `LattE` v1.1, 2003. Software package `LattE` is available at `http://www.math.ucdavis.edu/~latte/`
42. De Loera, J.A., Haws, D., Hemmecke, R., Huggins, P., Yoshida, R. A computational study of integer programming algorithms based on Barvinok's rational functions. *Disc. Optim.* **2**, 135–144 (2003)
43. De Loera, J.A., Hemmecke, R., Köppe, M., Weismantel, R. Integer polynomial optimization in fixed dimension. *Math. Oper. Res.* **31**, 147–153 (2006)
44. De Loera, J.A., Sturmfels, B. Algebraic unimodular counting. *Math. Progr.* **96**, 183–203 (2003)
45. Dyer, M.E. The complexity of vertex enumeration methods. *Math. Oper. Res.* **8**, 381–402 (1983)
46. Dyer, M.E., Frieze, A.M. The complexity of computing the volume of a polyhedron. *SIAM J. Comput.* **17**, 967–974 (1988)
47. Dyer, M.E., Frieze, A., Kannan, R. A random polynomial-time algorithm for approximating the volume of convex bodies. *J. ACM* **38**, 1–17 (1991)
48. Dyer, M.E., Frieze, A., Kannan, R., Kapoor, A., Perkovic, L., Vazirani, U. A mildly exponential time algorithm for approximating the number of solutions to a multi-dimensional knapsack problem. *Comb. Prob. Comput.* **2**, 271–284 (1993)
49. Dyer, M., Kannan, R. On Barvinok's algorithm for counting lattice points in fixed dimension, *Math. Oper. Res.* **22**, 545–549 (1997)
50. Dyer, M.E., Goldberg, L.A., Greenhill, C., Jerrum, M. The relative complexity of approximate counting problems. Approximation algorithms. *Algorithmica* **38** (2004), 471–500 (2004)
51. Ehrhart, E. Sur un problème de géométrie diophantienne linéaire II. *J. Reine Angew. Math.* **227**, 25–49 (1967)
52. Ehrhart, E. Sur les équations diophantiennes linéaires. *C. R. Acad. Sci., Paris, Sér. A* **288**, 785–787 (1979)

53. Eisenbrand, F. Fast integer programming in fixed dimension. In: G. Di Battista, U. Zwick, (eds.), Proceedings of the 11th Annual European Symposium on Algorithms, ESA' 2003, pp. 196–207. Lecture Notes in Computer Sciences **2832**, Springer Verlag, Berlin (2003)
54. Elekes, G. A geometric inequality and the complexity of measuring the volume. *Discr. Comp. Geom.* **1**, 289–292 (1986)
55. Erdös, P., Graham, R.L. On a linear diophantine problem of Frobenius. *Acta Arithm.* **21**, 399–408 (1972)
56. Filliman, P. The volume of duals and sections of polytopes. *Mathematika* **39**, 67–80 (1992)
57. Freiman, G. An analytical method of analysis of Boolean equations. *Ann. NY Acad. Sci.* **337**, 97–102 (1980)
58. Gomory, R.E. Some polyhedra related to combinatorial problems. *Lin. Alg. Appl.* **2**, 451–558 (1969)
59. Gomory, R.E., Johnson, E.L. The group problem and subadditive functions. In: T.C. Hu, S.M. Robinson, (eds.), Mathematical Programming. Academic Press, New York (1973)
60. Gomory, R.E., Johnson, E.L., Evans, L. Corner polyhedra and their connections with cutting planes, *Math. Program. Ser. B* **96**, 321–339 (2003)
61. Gomory, R.E., Johnson, E.L. T-space and cutting planes. Algebraic and geometric methods in discrete optimization. *Math. Program. Ser. B* **96**, 341–375 (2003)
62. Gritzmann, P., Klee, V. Basic problems in computational convexity II. In: T. Bisztriczky, P., McMullen, R., Schneider, A.I. Weiss, (eds.), Polytopes: Abstract, Convex and Computational. NATO ASI series, Kluwer Academic Publishers, Dordrecht (1994)
63. Güler, O. Barrier functions in interior point methods, Math. Oper. Res. **21**, 86–885 (1996)
64. Hiriart-Urruty, J.-B., Lemarechal, C. Convex Analysis and Minimization Algorithms I. Springer-Verlag, Berlin (1993)
65. Hosten, S., Sturmfels, B. Computing the integer programming gap, arXiv:math.OC/0301266v1, 2003.
66. Hosten, S., Thomas, R. Groebner bases and integer programming. In: B. Buchberger, F. Winkler (eds.), Groebner Bases and Applications, pp. 144–158, London Math. Soc. Lecture Notes Series **251**, Cambridge University Press, Cambridge (1998)
67. Hosten, S., Thomas, R. Gomory integer programs. *Math. Program.* **96**, 271–292 (2003)
68. Iosevich, A. Curvature, combinatorics, and the Fourier transform. Not. AMS **48**, 577–583 (2001)
69. Jeroslow, R. Cutting-plane theory: Algebraic methods. *Discr. Math.* **23**, 121–150 (978)
70. Jeroslow, R. Minimal inequalities. *Math. Program.* **17**, 1–15 (1979)
71. Jeroslow, R. An introduction to the theory of cutting-planes. *Ann. Discrete Math.* **5**, 71–95 (1979)
72. Jerrum, M., Sinclair, A. The Markov chain Monte Carlo method: An approach to approximate counting and integration. In: D.S. Hochbaum (ed.), Approximation Algorithms for NP-Hard Problems. PWS Publishing, Boston (1996)
73. Johnson, E.L. On the group problem and a subadditive approach to integer programming. *Ann. Discrete Math.* **5**, 97–112 (1979)
74. Johnson, E.L. Integer Programming: Facets, Subadditivity, and Duality for Group and Semi-group Problems. Society for Industrial and Applied Mathematics, Philadelphia (1980)
75. Kannan, R. Lattice translate of a polytope and the Frobenius problem. *Combinatorica* **12**, 161–177 (1992)
76. Kannan, R., Lovász, L., Simonovits, M. Random walks and an $O^*(n^5)$ volume algorithm for convex bodies. *Random Struct. Algo.* **11**, 1–50 (1997)
77. Kantor, J.-M., Khovanskii, A. Une application du théorème de Riemann–Roch combinatoire au polynôme d'Ehrhart des polytopes entiers. *C. R. Acad. Sci. Paris (Série I)* **317**, 501–507 (1993)
78. Khachyan, L. Complexity of polytope volume computation. In: J. Pach, (ed.), New Trends in Discrete and Computational Geometry, pp. 91–101. Springer-Verlag, Berlin (1993)
79. Khovanskii, A., Pukhlikov, A. A Riemann–Roch theorem for integrals and sums of quasipolynomials over virtual polytopes. *St–Petersburg Math. J.* **4**, 789–812 (1993)
80. Klabjan, D. A new subadditive approach to integer programming. In: W.J. Cook, (ed.), Integer Programming and Combinatorial Optimization, pp. 384–400. Springer Verlag, Berlin (2002)
81. Klabjan, D. A practical algorithm for computing the subadditive dual function for set-partitioning. *Comput. Optim. Appl.* **29**, 347–368 (2004)
82. Kollár, J. Sharp effective Nullstellensatz. *J. Am. Math. Soc.* **1**, 963–975 (1988)
83. Köppe, M., Louveaux, Q., Weismantel, R., Wolsey, L.A. Extended formulations for Gomory corner polyhedra. *Disc. Optim.* **1**, 141–165 (2004)
84. Kozlov, M.K. An approximate method of calculating the volume of a convex polyhedron. *U.S.S.R. Comput. Math. Math. Phys.* **22**, 227–233 (1982)

85. Oeuvres de Laguerre. Tome I. Algèbre. Calcul intégral. Rédigées par Ch. Hermite, H. Poincaré et E. Rouché. Réimpression de l'édition de (1898) Chelsea Publishing Co., Bronx, N.Y., (1972)
86. Lasserre, J.B. Integer programming duality and superadditive functions. Contemp. Math. **374**, 139–150 (2005)
87. Lasserre, J.B. Integer programming duality. *Proc. Symp. Appl. Math.* **61**, 67–83 (2004)
88. Lasserre, J.B. Duality and a Farkas lemma for integer programs, In: E. Hunt, C.E.M. Pearce, (eds.), Optimization: Structure and Applications. Kluwer Academic Publishers (2008)
89. Lasserre, J.B. Generating functions and duality for integer programs. *Discr. Optim.* **1**, 167–187 (2004)
90. Lasserre, J.B. A discrete Farkas lemma. *Discr. Optim.* **1**, 67–75 (2004)
91. Lasserre, J.B. The integer hull of a convex rational polytope. *Discr. Comp. Geom.* **32**, 129–139 (2004)
92. Lasserre, J.B. Integer programming, Barvinok's counting algorithm and Gomory relaxations. *Oper. Res. Letters* **32**, 133–137 (2003)
93. Lasserre, J.B. La valeur optimale des programmes entiers. *C. R. Acad. Sci. Paris. Ser. I* **335**, 863–866 (2002)
94. Lasserre, J.B. An analytical expression and an algorithm for the volume of a convex polyhedron in $\mathbb{R}^n$. *J. Optim. Theor. Appl.* **39**, 363–377 (1983)
95. Lasserre, J.B., Zeron, E.S. A Laplace transform algorithm for the volume of a convex polytope. *J. ACM* **48**, 1126–1140 (2001)
96. Lasserre, J.B., Zeron, E.S. Solving a class of multivariate integration problems via Laplace techniques. *Appl. Math. (Warsaw)* **28**, 391–405 (2001)
97. Lasserre, J.B., Zeron, E.S. Solving the knapsack problem via $\mathbb{Z}$-transform. *Oper. Res. Lett.* **30**, 394–400 (2002)
98. Lasserre, J.B., Zeron, E.S. On counting integral points in a convex rational polytope. *Math. Oper. Res.* **28**, 853–870 (2003)
99. Lasserre, J.B., Zeron, E.S. An alternative algorithm for counting lattice points in a convex polytope. *Math. Oper. Res.* **30**, 597–614 (2005)
100. Lasserre, J.B., Zeron, E.S. Simple explicit formula for counting lattice points of polyhedra. In: M. Fischetti, D.P. Williamson (eds.), *Integer Programming and Combinatorial Optimization*, pp. 367–381 Springer-Verlag, Berlin (2007)
101. Laurent, M. A comparison of the Sherali–Adams, Lovász–Schrijver and Lasserre relaxations for 0-1 programming. *Math. Oper. Res.* **30**, 470–496 (2003)
102. Lawrence, J. Polytope volume computatio. *Math. Comp.* **57**, 259–271 (1991)
103. Lenstra, H.W. Integer programming with a fixed number of variables. *Math. Oper. Res.* **8**, 538–548 (1983)
104. Lenstra, A.K., Lenstra, H.W., Lovász, L. Factoring polynomials with rational coefficients, *Math. Ann.* **261**, 515–534 (1982)
105. Llewellyn, D., Ryan, J. A primal dual integer programming algorithm. *Discr. Appl. Math.* **45**, 261–275 (1993)
106. Litvinov, G.L., Maslov, V.P., Shpiz, G.B. Linear functionals on idempotent spaces: An algebraic approach. *Doklady Math.* **58**, 389–391 (1998)
107. Lovász, L. An Algorithmic Theory of Numbers, Graphs and Convexity. SIAM, Philadelphia (1986)
108. Lovász, L., Scarf, H.E. The generalized basis reduction algorithm, *Math. Oper. Res.* **17**, 751–764 (1992)
109. Marchand, H., Martin, A., Weismantel, R., Wolsey, L.A. Cutting planes in integer and mixed integer programming. *Disc. Appl. Math.* **123**, 397–446 (2002)
110. Mayr, E.W., Meyer, A.R. The complexity of the word problems for commutative semigroups and polynomial ideals. *Adv. Math.* **46**, 305–329 (1982)
111. Mitrinović, D.S., Sándor, J., Crstici, B. Handbook of Number Theory. Kluwer Academic Publishers, Dordrecht (1996)
112. Nemhauser, G.L. Duality for Integer Optimization. In: M., O'hEigeartaigh, J.K., Lenstra, A.H.G. Rinnooy Kan, (eds.), Combinatorial Optimization: Annotated Bibliographies. John Wiley & Sons, Chichester, pp. 11–20 (1985)
113. Nemhauser, G.L., Wolsey, L.A. Integer and Combinatorial Optimization. John Wiley & Sons, city (1988)
114. Nesterov, Y. Fast Fourier transform and its application to integer knapsack problems. CORE discussion paper 2004/64, Core, Université Louvain-la-Neuve, Belgium, New York (2004)
115. Netto, E. Lehrbuch der Combinatorik. B.G. Teubner, Leipzing and Berlin, 2nd ed. (1927)
116. Niederreitter, H. Random Number Generation and Quasi-Monte Carlo Methods. Society for Industrial and Applied Mathematics, Philadelphia, Pennsylvania (1992)
117. Rockafellar, R.T. Convex Analysis. Princeton University Press, Princeton (1972)
118. Ryan, J. The structure of an integral monoid and integer programming feasibility. *Discr. Appl. Math.* **28**, 251–263 (1990)

119. Ryan, J. The constraints of the group of an integral monoid. *Lin. Alg. Appl.* **139**, 285–292 (1990)
120. Ryan, J., Trotter, L.E. Weyl–Minkowski duality for integral monoids. *Optimization* **28**, 315–331 (1994)
121. Schrijver, A.J. Theory of Linear and Integer Programming. John Wiley & Sons, Chichester (1986)
122. Seidenber, A. Constructions in algebra. *Trans. Am. Math. Soc.* **197**, 273–313 (1974)
123. Selmer, S. On a linear diophantine problem of Frobenius. *J. Reine Angew.* Math. **293/294**, 1–17 (1977)
124. Shapiro, J.F. Convergent duality for the travelling salesman problem. *Oper. Res. Lett.* **10**, 129–136 (1991)
125. Shapiro, J.F. Mathematical Programming: Structures and Algorithms. John Wiley & Sons, New York (1979)
126. Sturmfels, B., Thomas, R.R. Variation of cost functions in integer programming. *Math. Program.* **77**, 357–387 (1997)
127. Thomas, R.R. A geometric Buchberger algorithm for integer programming. *Math. Oper. Res.* **20**, 864–884 (1995)
128. Thomas, R.R. Algebraic methods in integer programming. In: C., Floudas, P. Pardalos, (eds.), Encyclopedia of Optimization. Kluwer Academic Publishers, Dordrecht, pp. 499–509 (2001)
129. Tind, D., Wolsey, L.A. An elementary survey of general duality theory in mathematical programming. *Math. Program.* **21**, 241–261 (1981)
130. Truong, V.A., Tunçel, L. Geometry of homogeneous convex cones, duality mapping, and optimal self-concordant barriers. Research report COOR #2002-15, University of Waterloo, Waterloo, Canada (2002)
131. Von Hohenbalken, B. Finding simplicial subdivisions of polytopes. *Math. Program.* **21**, 233–234 (1981)
132. Woods, K., Yoshida, R. Short rational generating functions and their applications to integer programming. SIAG/OPT Views-and-News **16**, 15–19 (2005)
133. Wolsey, L.A. Extensions of the group theoretic approach in integer programming. *Manag. Sci.* **18**, 74–83 (1971)
134. Wolsey, L.A. Integer programming duality: Price functions and sensitivity analysis. *Math. Program.* **20**, 173–195 (1981)
135. Wolsey, L.A. The b-hull of an integer program. *Disc. Appl. Math.* **3**, 193–201 (1981)
136. Wolsey, L.A. Integer Programming. John Wiley & Sons, Inc., New York (1998)

Glossary

$\mathbb{N}$— set of natural numbers

$\mathbb{Z}$— set of integers

$\mathbb{Q}$— set of rational numbers

$\mathbb{R}$— set of real numbers

$\mathbb{C}$— set of complex numbers

A— matrix in $\mathbb{R}^{m\times n}$, or $\mathbb{Z}^{m\times n}$, or $\mathbb{N}^{m\times n}$

A_j— column j of matrix A

$\Re(z)$— real part of $z\in\mathbb{C}$, or vector of real parts of $z\in\mathbb{C}^n$

$|z|$— modulus of $z\in\mathbb{C}$, or vector of moduli of $z\in\mathbb{C}^n$

$\Omega(y)$— convex polyhedron $\{x\in\mathbb{R}^n\,|Ax=y,x\geq 0\}$

$\mathrm{conv}\,(\Omega(y)\cap\mathbb{Z}^n)$— integer hull of $\Omega(y)$

$\mathbf{P}$— linear program $\max\{c'x\,|x\in\Omega(y)\}$

$\mathbf{P}^*$— LP dual of $\mathbf{P}$

$\mathbf{P}_d$— integer program $\max\{c'x\,|x\in\Omega(y)\cap\mathbb{Z}^n\}$

$\mathbf{P}_d^*$— dual problem of $\mathbf{P}_d$

$\mathbf{I}$— integration program $\int_{\Omega(y)}\mathrm{e}^{c'x}\,d\sigma$

$\mathbf{I}^*$— dual of $\mathbf{I}$ (a complex integral)

$\mathbf{I}_d$— summation program $\sum_{x\in\Omega(y)\cap\mathbb{Z}^n}\mathrm{e}^{c'x}$

$\mathbf{I}_d^*$— dual of $\mathbf{I}_d$ (a complex integral)

$f(y,c),f_d(y,c)$— respective optimal values of $\mathbf{P}$ and $\mathbf{P}_d$

$\widehat{f}(y,c),\widehat{f}_d(y,c)$— respective values of $\mathbf{I}$ and $\mathbf{I}_d$

$F(\lambda,c)$— Fenchel transform: $\lambda \mapsto \inf_{y\in\mathbb{R}^m} \lambda' y - f(y,c)$ (as $f(\cdot,c)$ is concave)

$\widehat{F}(\lambda,c)$— Laplace transform: $\lambda \mapsto \int_{\mathbb{R}^m} e^{\lambda' y} f(y,c)\,dy$

$\widehat{F}_d(z,c)$— $\mathbb{Z}$-transform: $z \mapsto \sum_{y\in\mathbb{Z}^m} z^{-y} f_d(y,c)$

$\ln z$— logarithm (Neperian) of $z \in \mathbb{C}$

$\mathbb{R}[z_1,\ldots,z_m]$— ring of polynomials in m variables, with coefficients in $\mathbb{R}$

z^α: with $z \in \mathbb{C}^m, \alpha \in \mathbb{Z}^m$: Laurent monomial $z_1^{\alpha_1}\cdots z_m^{\alpha_m}$

z^{A_j}: with $z \in \mathbb{C}^m, A_j \in \mathbb{Z}^m$: Laurent monomial $z_1^{A_{1j}}\cdots z_m^{A_{mj}}$

$\oplus$— depending on the underlying algebra, stands for either the usual addition $a\oplus b = a+b$ or $a\oplus b = \max[a,b]$

Δ— is $\mathbb{R}^n$ (resp., $\mathbb{Z}^n$) in the continuous (resp., discrete) context

$\int^\oplus$— depending on Δ and the underlying algebra, stands for $\int$ or $\sum$ or max

Index

Printed in the United States of America